Springer Theses

Recognizing Outstanding Ph.D. Research

For further volumes:
http://www.springer.com/series/8790

Aims and Scope

The series "Springer Theses" brings together a selection of the very best Ph.D. theses from around the world and across the physical sciences. Nominated and endorsed by two recognized specialists, each published volume has been selected for its scientific excellence and the high impact of its contents for the pertinent field of research. For greater accessibility to non-specialists, the published versions include an extended introduction, as well as a foreword by the student's supervisor explaining the special relevance of the work for the field. As a whole, the series will provide a valuable resource both for newcomers to the research fields described, and for other scientists seeking detailed background information on special questions. Finally, it provides an accredited documentation of the valuable contributions made by today's younger generation of scientists.

Theses are accepted into the series by invited nomination only and must fulfill all of the following criteria

- They must be written in good English.
- The topic should fall within the confines of Chemistry, Physics, Earth Sciences, Engineering and related interdisciplinary fields such as Materials, Nanoscience, Chemical Engineering, Complex Systems and Biophysics.
- The work reported in the thesis must represent a significant scientific advance.
- If the thesis includes previously published material, permission to reproduce this must be gained from the respective copyright holder.
- They must have been examined and passed during the 12 months prior to nomination.
- Each thesis should include a foreword by the supervisor outlining the significance of its content.
- The theses should have a clearly defined structure including an introduction accessible to scientists not expert in that particular field.

Akihiko Monnai

Relativistic Dissipative Hydrodynamic Description of the Quark-Gluon Plasma

Doctoral Thesis accepted by
The University of Tokyo, Tokyo, Japan

Springer

Author (Current Address)
Dr. Akihiko Monnai
RIKEN BNL Research Center
New York
USA

Supervisor
Prof. Tetsuo Hatsuda
The University of Tokyo
Tokyo
Japan

ISSN 2190-5053 ISSN 2190-5061 (electronic)
ISBN 978-4-431-54797-6 ISBN 978-4-431-54798-3 (eBook)
DOI 10.1007/978-4-431-54798-3
Springer Tokyo Heidelberg New York Dordrecht London

Library of Congress Control Number: 2013957879

Printed on acid-free paper

Springer is part of Springer Science+Business Media (www.springer.com)

Parts of this thesis have been published in the journal articles:

- A. Monnai, "Dissipative hydrodynamic effects on baryon stopping", Phys. Rev. C **86**, 014908 (2012).
- A. Monnai and T. Hirano, "Longitudinal viscous hydrodynamic evolution for the shattered colour glass condensate", Phys. Lett. B **703**, 583 (2011).
- A. Monnai and T. Hirano, "Relativistic dissipative hydrodynamic equations at the second order for multi-component systems with multiple conserved currents", Nucl. Phys. A **847**, 283 (2010).
- A. Monnai and T. Hirano, "Effects of bulk viscosity at freezeout", Phys. Rev. C **80**, 054906 (2009).

Supervisor's Foreword

In the early Universe at temperatures greater than 10^{12} K, the ordinary matter composed of protons and neutrons was dissolved into a plasma of elementary particles, the quarks and the gluons. Such a primordial state of matter is called the quark-gluon plasma (QGP), and its properties are currently under active investigation by using the relativistic heavy-ion collisions at the Relativistic Heavy Ion Collider (RHIC) and at the Large Hadron Collider (LHC).

The theoretical basis for studying QGP is provided by the color SU(3) gauge theory of strong interaction, the quantum chromodynamics (QCD). To make a connection between QCD predictions and experimental data, it is essential to formulate a general framework of describing the space–time evolution of hot/dense matter produced in relativistic heavy-ion collisions. The relativistic hydrodynamics supplemented with suitable initial conditions is known be one such framework. In recent years, the importance of considering dissipative effects beyond the assumption of perfect QGP fluid has been recognized both theoretically and experimentally.

The present thesis by Dr. Akihiko Monnai introduces a consistent formalism of the relativistic dissipative hydrodynamics with multiple conserved charges for the first time. The second-order dissipative hydrodynamic equations obtained from this formalism lead naturally to the various cross-coupling effects. Also, these equations with a new numerical algorithm are applied to study the effect of secondary interactions in the viscous medium created at RHIC and LHC. Furthermore, by using these equations with net baryon density, influences of the viscosity and baryon diffusion on the baryon stopping in relativistic heavy-ion collisions are investigated.

The approach developed by Dr. Monnai offers a theoretical framework not only for making a firm bridge between QCD and heavy-ion experiments but also for a deeper understanding of strongly interacting fluid created at RHIC and LHC. Also, the present thesis contains a transparent summary of the current status of the theoretical and experimental studies of QGP, making it a good introductory monograph for students and researchers who are interested in this field.

It is with the greatest pleasure that I introduce Dr. Akihiko Monnai's work for publication in the Springer Theses series. His work was nominated as an outstanding Physics Ph.D. Thesis of Fiscal Year 2012 by the Department of Physics, Graduate School of Science, The University of Tokyo.

Tokyo, November 2013 Tetsuo Hatsuda

Acknowledgments

I would like to thank my advisor, Prof. Tetsuo Hatsuda, for insightful guidance, helpful support, great inspiration, and enlightening discussion. This Ph.D. thesis was written under his supervision.

It is my great pleasure to acknowledge that I had a pleasant life as a Ph.D. student at the University of Tokyo with an excellent environment for research. I would especially like to thank my past and current colleagues Dr. A. Rothkopf, Dr. Y. Akamatsu, and Mr. Y. Araki. I am also grateful to RIKEN, which provided me with wonderful opportunities to attend scientific meetings and to meet researchers in various academic fields in my final year.

My great gratitude goes to Prof. B. Mueller of Duke University in Durham NC, USA, for inspiring and valuable discussion. I also appreciate Prof. S. Bass and Dr. H. Petersen for their hospitality during my 2 month stay at Duke University.

It should be mentioned that Prof. T. Hirano was my advisor in my first year, before his departure. I hereby express my gratitude to him for introducing me to heavy ion physics.

I would like to thank Prof. H. Fujii, Prof. K. Itakura, Prof. T. Kodama, Prof. T. Koide, Prof. T. Kunihiro, Prof. T. Matsui, Prof. S. Muroya, Prof. Y. Nara, Prof. C. Nonaka, Prof. A. Ohnishi, and Dr. G. S. Denicol for their valuable and stimulating comments and discussion.

Finally, I personally thank my family and friends who supported me and provided me with peace of mind. My work was supported by Grant-in-Aid for JSPS Research Fellows (No. 10J07647) for the entire 3 years of my Ph.D. course.

Contents

Abbreviations

Acronyms

AdS	Anti-de Sitter
AGS	Alternating Gradient Synchrotron
BES	Beam energy scan
CFT	Conformal field theory
CGC	Color glass condensate
FAIR	Facility for Antiprotons and Ion Research
HERA	Hadron Electron Ring Accelerator
J-PARC	Japan Proton Accelerator Research Complex
KLN	Kharzeev-Levin-Nardi
LHC	Large Hadron Collider
LRF	Local rest frame
MC	Monte-Carlo
NICA	Nuclotron-based Ion Collider Facility
pQCD	Perturbative QCD
RHIC	Relativistic Heavy Ion Collider
SIS	Heavy-ion synchrotron
SPS	Super Proton Synchrotron
sQGP	Strongly coupled QGP
QCD	Quantum chromodynamics
QED	Quantum electrodynamics
QGP	Quark-gluon plasma

Symbols of Nuclei

p	Proton
d	Deuteron
Cu	Copper
Au	Gold
Pb	Lead
U	Uranium

Quantum Chromodynamics

α_s	QCD running coupling
N_c	Number of colors
N_f	Number of flavors
Λ_{QCD}	QCD scale parameter
T_c	Crossover temperature

Heavy Ion Collisions

$\sqrt{s_{NN}}$	Center-of-mass energy
y_p	Beam rapidity of incoming projectile
S_{area}	Overlapping area
N	Particle number
N_{ch}	Charged particle number
$N_{\mathrm{B}-\bar{\mathrm{B}}}$	Net baryon number
N_{coll}	Number of collisions
N_{part}	Number of participants
E_T	Transverse energy
b	Impact parameter
v_2	Elliptic flow
v_n	n-th order flow harmonics
Ψ	Reaction plane
R_{AA}	Nuclear modification factor

Coordinates

τ	Proper time
η_s	Space-time rapidity
r	Radius
ϕ	Angle
m_T	Transverse mass
y	Rapidity
p_T	Transverse momentum
η_p	Pseudo-rapidity
ϕ_p	Angle in momentum space

Hydrodynamics

$g^{\mu\nu} = \mathrm{diag}(+,-,-,-)$	Minkowski metric
u^μ	Flow
$\Delta^{\mu\nu} = g^{\mu\nu} - u^\mu u^\nu$	Projection operator
$D = u^\mu \partial_\mu$	Time-like derivative

$\nabla^\mu = \Delta^{\mu\nu}\partial_\nu$	Space-like derivative
Y_f	Flow rapidity
$T^{\mu\nu}$	Energy-momentum tensor
e_0	Energy density
P_0	Hydrostatic pressure
Π	Bulk pressure
c_s	Sound velocity
W^μ	Energy dissipation current
$\pi^{\mu\nu}$	Shear stress tensor
N_J^μ	J-th conserved charge current
n_{J0}	J-th conserved charge density
V_J	J-th conserved charge dissipation current
s^μ	Entropy current
s	Entropy density
T	Temperature
μ_J	Chemical potential of the J-th conserved charge
η	Shear viscosity
ζ	Bulk viscosity
κ_W	Thermal conductivity
κ_{V_J}	Charge conductivity of the J-th net charge
D_{V_J}	Diffusion coefficient of the J-th net charge
$\tau_\Pi, \tau_W, \tau_{V_J}, \tau_\pi$	Relaxation times
χ	Second-order transport coefficients
$\theta = \nabla_\mu u^\mu$	Expansion scalar
$\sigma^{\mu\nu} = \nabla^{\langle\mu}u^{\nu\rangle}$	Traceless symmetric thermodynamic force (see "Tensor operations" below for the definition)
$\omega^{\mu\nu} = \nabla^\mu u^\nu - \nabla^\nu u^\mu$	Vorticity

Tensor Operations

$A^{\mu\nu} = A_{\parallel\parallel}u^\mu u^\nu + \frac{1}{3}A_{\perp\mathrm{T}}\Delta^{\mu\nu} + A^\mu_{\perp\parallel}u^\nu + A^\nu_{\perp\parallel}u^\mu + A^{\mu\nu}_{\perp\perp}$	Decomposition (rank-2)
$A^\mu = A_\parallel u^\mu + A^\mu_\perp$	Decomposition (rank-1)
$A^{(\mu}B^{\nu)} = \frac{1}{2}(A^\mu B^\nu + A^\nu B^\mu)$	Symmetrization
$A^{\langle\mu}B^{\nu\rangle} = \left[\Delta^\mu_{(\alpha}\Delta^\nu_{\beta)} - \frac{1}{3}(\Delta^{\mu\nu}\Delta_{\alpha\beta})\right]A^\alpha B^\beta$	Traceless symmetrization

Kinetic Theory

f^i	Phase-space distribution (of i-th particle species)
f_0^i	Local equilibrium distribution
$\delta f^i = f^i - f_0^i$	Off-equilibrium distortion of the distribution
y^i	Exponent in the distribution
y_0^i	Exponent in the local equilibrium distribution

$\delta y^i = y^i - y^i_0$	Off-equilibrium distortion of y^i
ε	Quantum statistics (bosons: +1, fermions: −1, classical particles: 0)
q^i_J	Quantum number of the J-th conserved charge
b_i	Baryon number
g_i	Degeneracy
$\varepsilon_{\mu\nu}, \varepsilon^J_\mu$	Distortion tensors in momentum expansion of δf
$I^{\mu_1 \cdots \mu_n}$	n-th moment of f^i
$J^{\mu_1 \cdots \mu_n}, K^{\mu_1 \cdots \mu_n}$	n-th moment of f^i_0 with quantum statistics (see Appendix C)

Color Glass Condensate

Bjorken-x	Momentum fraction of incident particles
Q_s	Saturation scale
λ	Parameter for rapidity dependence
φ	Unintegrated gluon distribution

Freeze-out

T_f	Freeze-out temperature
Σ	Freeze-out hypersurface
$d\sigma^\mu$	Freeze-out hypersurface element

Curriculum Vitae

Akihiko Monnai

Current Affiliation:
RIKEN-BNL Research Center
Building 510A, Physics Department, Brookhaven National Laboratory, Upton, NY 11973-5000, USA
E-mail: amonnai@riken.jp
Web: http://tkynt2.phys.s.u-tokyo.ac.jp/monnai/
Former Affiliation:
Department of Physics, Graduate School of Science,
The University of Tokyo, Japan
7-3-1 Hongo, Bunkyo-ku, Tokyo 113-0033, Japan

Employment

July 2013—present, RIKEN-BNL Research Center, Nishina Center for Accelerator-Based Science, RIKEN
Special postdoctoral researcher

April 2013—June 2013, Theoretical Research Division, Nishina Center for Accelerator-Based Science, RIKEN
Special postdoctoral researcher

Education/Degrees

April 2010—March 2013, Doctor of Science (Ph.D.)
Department of Physics, Graduate School of Science, The University of Tokyo
(Advisor: Prof. Tetsuo Hatsuda)

April 2008—March 2010, Master of Science (M.Sc.)
Department of Physics, Graduate School of Science, The University of Tokyo
(Advisor: Prof. Tetsufumi Hirano)

April 2004—March 2008, Bachelor of Science (B.S.)
Department of Physics, Faculty of Science, The University of Tokyo

Appointments

April 2010—March 2013, Research Fellow (DC1) of the Japanese Society for the Promotion of Science

Awards

March 2013, Research Award of Graduate School of Science, the University of Tokyo for Doctor Course

June 2010, "Most Delicious Talk" Award, Hot Quarks 2010 Organizing Committee

March 2007, Dean Award of Faculty of Science, the University of Tokyo for Elite Science Student Visit Abroad Program

Community Services

Referee to Journal of Physics G: Nuclear and Particle Physics (Institute of Physics)

Teaching Experiences

April 2009—March 2010 Teaching Assistant for Undergraduate Class "Computational Experiment" at Department of Physics, the University of Tokyo

April 2008—September 2008 Teaching Assistant for Undergraduate Class "Nuclear Physics" at Department of Physics, the University of Tokyo

Research Statement

My fields of research include theoretical studies of non-equilibrium collective dynamics in the quark-gluon plasma (QGP). The QGP is a many-body system of elementary particles governed by quantum chromodynamics that appears at very high temperatures above two trillion degrees. It is considered to once have filled the early universe just after the Big Bang. High-energy heavy ion collisions have succeeded in producing the hot matter on Earth and revealed it to behave as a relativistic fluid with small yet non-vanishing viscosity. My researches so far include formulation of causal relativistic dissipative hydrodynamic theory and its application to numerical analyses of the QGP in collider experiments at finite densities. I aim to explicate the macroscopic time-evolution of the hot medium as well as the microscopic properties such as equation of state and transport coefficients through comparing theoretical predictions and experimental data.

Chapter 1
Introduction

Our approach to physics seems to be based on two distinct perspectives: reductionism and holism. The former states physical phenomena can be understood from the fundamental building blocks. In current understandings, this statement means that all physics should arise from the fields of elementary particles and the dynamics which governs them. There are three known generations of quarks and leptons, each hosting a weak isospin doublet: up and down, strange and charm, bottom and top as quarks, along with electron and electron neutrino, muon and muon neutrino and tau and tau neutrino as leptons. They are fermions and have antiparticle partners. On the other hand, there are five gauge bosons discovered so far: photon γ, weak bosons $W^\pm$ and Z^0, and gluon g which convey the electroweak and the strong forces. The electroweak theory of the standard model predicts the existence of the Higgs boson H^0 for mass generation mechanism for the weak bosons, hints of which are now observed in a hadronic collider experiment. It is subsequently considered to be the origin of mass for the quarks and the leptons by introducing the Yukawa interactions between the fields. The gravity is the only one of the four fundamental forces that has so far been out of the standard model of particle physics, due to the lack of proper renormalization method for taming its quantum divergence. It is theorized to require the spin-2 boson called graviton G corresponding to the fact that the energy-momentum tensor is rank-2. Note that Higgs mechanism does not produce the gravitational mass but the inertial mass. General relativity states the inertial mass and the gravitational mass are equivalent, and is currently beyond the scope of quantum dynamics. Various theoretical studies such as super string theories are being investigated to describe quantum gravity.

Of the four fundamental forces the strong and the weak forces are short-ranged unlike the other two. The latter is the case because it is conveyed by the massive weak boson fields. The strong force, on the other hand, is a short range force even though the gluon is massless, because the color charge is completely confined in color-neutral composite particles in vacuum. As a result at nuclear energy scale the strong force appears between the nucleons in terms of massive pions, limiting the range of the interaction. It can be seen that the strong interaction would be more powerful than electromagnetic interaction since it binds the nucleons together against the Coulomb repulsion of protons in nuclei.

A. Monnai, *Relativistic Dissipative Hydrodynamic Description of the Quark-Gluon Plasma*, Springer Theses, DOI: 10.1007/978-4-431-54798-3_1,

The holism, on the other hand, emphasizes the importance of collective physics of many-body systems and offers phenomenological description. The microscopic features are integrated out to describe specific aspects behind the observed phenomena with limited number of variables. It does not necessarily require equilibration of the medium, though the collective description of highly-chaotic systems are generally very difficult. When the system is reasonably close to equilibrium, statistical thermodynamics becomes a powerful method to describe the physics. One of the most distinctive properties of the macroscopic theories is that they are time-irreversible unless they are specifically idealized to be otherwise, i.e., the time-arrow appears in the theory. This fact is represented by the law of increasing entropy.

These two concepts are seemingly opposite in nature and are often perceived as mutually exclusive, as the famous quotations go "the rest is chemistry" for reductionism and "more is different" for holism. However this should not be always the case, since the two are often connected at more fundamental level. The concept of spontaneous symmetry breaking, for example, explains the mass generation mechanism in particle physics as well as the magnetization of a ferromagnet or the generation of quantum vortices for a Bose-Einstein condensate in statistical mechanics. Also it is naturally expected that one has to focus on both microscopic and macroscopic aspects to understand collective dynamics of elementary particles. This encourages one to construct an interdisciplinary model by bridging particle physics, nuclear physics and non-equilibrium statistical physics. In this thesis, I would like to show that a many-body system of quarks and gluons are described by the relativistic version of dissipative hydrodynamics at high energies.

1.1 Quantum Chromodynamics

Quantum chromodynamics (QCD) [1] is the fundamental description of quarks and gluons under the strong interaction. The dynamics of fields is given by the Euler-Lagrange equations. Since the strong interaction only affects quarks and gluons, the QCD Lagrangian reads

$$\mathscr{L}_S = \bar{q}_f^\alpha i\gamma^\mu(\partial_\mu - igA_\mu^{\ a}T^a)^{\alpha\beta} q_f^\beta - m_f\delta^{\alpha\beta}\bar{q}_f^\alpha q_f^\beta - \frac{1}{4}F_{\mu\nu}^a F^{a\mu\nu} + \mathscr{L}_g, \quad (1.1)$$

where q_α^f is the quark field with $f = u, d, s, c, b, t$ as the flavor index and $\alpha = 1, 2, 3$ as the index for the color triplet. A_a^μ is the gluon field where $a = 1, ..., 8$ is the index for the color octet. T^a is the fundamental representation of $SU(3)_C$ algebra which satisfies $[T^a, T^b] = if_{abc}T^c$. g is the QCD coupling constant and m_f is the quark mass. The field strength is $F_{\mu\nu}^a = \partial_\mu A_\nu^a - \partial_\nu A_\mu^a + gf_{abc}A_\mu^b A_\nu^c$. $\mathscr{L}_g$ represents the gauge-fixing term and the Faddeev-Popov term required for the quantum Lagrangian. It is also important to note that when quantum anomaly is neglected the Lagrangian has chiral symmetry in the massless limit, i.e., it is invariant under $U_L(N_f) \times U_R(N_f)$ transformation.

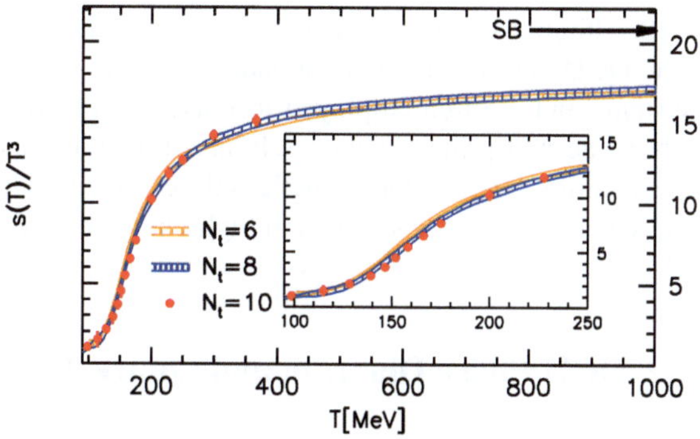

Fig. 1.1 The entropy density at vanishing chemical potential as a function of the temperature in a lattice QCD calculation. N_τ is the temporal extent which is related with lattice spacing. SB denotes the Stefan-Boltzmann limit. (Figure from Ref. [4], Springer Science+Business Media.)

The theory is very similar to quantum electrodynamics (QED) except that QCD has $SU(3)_C$ symmetry instead of $U(1)_{EM}$ symmetry. Unlike the photon in QED, the gluons themselves carry charges. As a result QCD exhibits a very unique behavior called asymptotic freedom [2, 3]. If one considers the renormalization of QCD, the running coupling $\alpha_s = g/4\pi$ at the lowest order is expressed as

$$\alpha_s(\mu) = \frac{1}{4\pi\beta_0 \ln(\mu^2/\Lambda_{\rm QCD}^2)}\left[1 - \frac{\beta_1}{\beta_0^2}\frac{\ln(\ln(\mu^2/\Lambda_{\rm QCD}^2))}{\mu^2/\Lambda_{\rm QCD}^2} + ...\right], \tag{1.2}$$

where $\beta_0 = [11 - (2/3)N_f]/(4\pi)^2$ and $\beta_1 = [102 - (38/3)N_f]/(4\pi)^2$ stay positive for QCD. μ is the energy where the divergence is renormalized and $\Lambda_{\rm QCD}$ is the QCD scale parameter. This suggests that the coupling becomes weaker as μ increases. On the other hand, the QCD medium can be strongly-coupled and highly non-perturbative at lower energies.

It is possible to perform perturbative calculations for a QCD medium when the energy is high and the coupling is small enough. However, when the coupling is strong, which turns out to be the case for the quark-gluon medium around the pseudo-critical temperature in the next chapter, this is not the case. There one can perform non-perturbative simulations by discretizing the space-time into lattice. This is called lattice QCD and it provides a variety of information on the strongly-coupled QCD systems. The equation of state, which gives the relation among the thermodynamic variables in media, at vanishing chemical potential from a latest $(2+1)$-flavor lattice QCD [4] (Fig. 1.1) shows that quark-hadron transition is a crossover rather than a phase transition. The cross over temperature is $T_c \sim 170\,\text{MeV}$. It shows the sudden rise in the dimensionless entropy density. This is considered to embody the emergence of the quark-gluon degrees of freedom out of the hadronic ones.

One of the recent developments in the study of strongly-coupled gauge theories is anti-de Sitter/conformal field theory (AdS/CFT) correspondence [5], in which one asserts the $(d+1)$-dimensional classical gravity theory on anti-de Sitter space is dual to d-dimensional quantum gauge theory. It is originally meant for the correspondence to $\mathcal{N} = 4$ super Yang-Mills theory, but now is applied for non-conformal theories as well. Although the correspondence is not mathematically proven nor it describes realistic QCD systems, the correspondence allows one to investigate strongly-coupled systems and is often employed to make conjectures for the quark-gluon medium where the first-principle approach has difficulties, such as dynamical responses of the medium. One notable example is the conjectured lower bound for shear viscosity [6] which seems to be satisfied in high-energy heavy ion collisions.

1.2 QCD Phase Diagram and Quark-Gluon Plasma

The quark matter under the strong interaction is known to have various different phases because of the unique natures of the QCD (Fig. 1.2 [7]). Due to the complexities in the first principle calculations at finite temperature and chemical potential, its phase diagram, however, is still not fully known theoretically. Thus it is one of the most important goals in high-energy physics to explicate it and uncover the properties of the hot and dense matter. So far various theories are proposed for determination of properties of the QCD matter.

It is known that the hadron is the most stable form of the QCD matter when the temperature and the chemical potential is small. The color charge—and thus bare quark—is confined in hadrons. Phenomenologically speaking, this implies that it becomes energetically more favorable to create a quark-antiquark pair than exposing bare quarks to the vacuum. Although there is no rigorous proof for the confinement, numerical estimations have shown a clear sign of it [8]. Thus despite its similarity as a charge-neutral composite particle to atoms, the hadrons do not ionize with color charge near vacuum. All the composite particles of quarks discovered so far are baryons, which has three quarks in different colors, and mesons, which is a quark-antiquark system with the opposite colors. There are so far no firm experimental evidences for other color neutral systems such as pentaquarks or dibaryons, or quark-less glue balls. It is an intriguing issue for theoretical nuclear physics to find if those states are allowed or not.

Liquid-gas transition takes place at the vanishing temperature limit when the density rises up and reaches $n \sim 0.17\,\mathrm{fm}^{-3}$. At this point hadrons form nuclear matter rather than separate clusters. This transition line meets an end point at a finite temperature.

At ultra-high temperature with low baryon chemical potential, the quark matter is more stable as the *quark-gluon plasma* (QGP), a phase of deconfined quarks and gluons, than as an ordinary nuclear matter [9]. Lattice QCD calculations suggest that the boundary between the QGP and the hadron phases is a crossover as mentioned earlier. The QGP is considered to have filled the early universe just after the Big Bang. The universe is expected to have experienced the electroweak phase transition after $t \sim 10^{-11}$ sec where the $\mathrm{SU}(2)_\mathrm{L} \times \mathrm{U}(1)_\mathrm{Y}$ symmetry of the theory had broken into the $\mathrm{U}(1)_\mathrm{EM}$ symmetry of the electromagnetic field, introducing small matter-antimatter asymmetry in the universe. The succeeding quark-gluon plasma phase is considered to persist until the QCD phase transition at $t \sim 10^{-5} - 10^{-4}$ s.

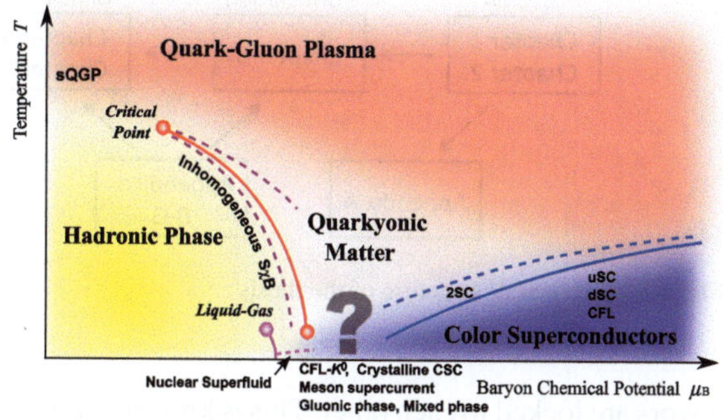

Fig. 1.2 A temperature-baryon chemical potential phase diagram of the hot and dense QCD matter. (Figure from Ref. [7], IOP Publishing.)

The nuclear collider experiment is an alternative place to find the QGP. It is believed that the deconfined matter is produced in heavy ion collisions at the Relativistic Heavy Ion Collider and the Large Hadron Collider (Chap. 2). The typical life-time of the man-made QGP is $t \sim 10^{-23}$ s and the expected temperature reaches 4×10^{12} K.[1] Due to its similarity to the early universe, heavy ion collisions are often referred to as *little bang*. So far various studies have been done to catch a glimpse of the early universe. It is found in the collider experiments that the QGP would be strongly coupled in the vicinity of the quark-hadron transition. The phenomenological description of the quark-gluon plasma in relativistic dissipative hydrodynamics will be the main focus of this thesis from the Chap. 2.

At very high temperature and density, the chiral symmetry is restored for quarks. The current quark masses are of $\mathcal{O}(1)$ MeV. At low temperature and density, a spontaneous symmetry breaking takes place with the chiral condensate $\langle \bar{q}q \rangle$ as the order parameter, through which quarks gain the mass of $\mathcal{O}(10^3)$ MeV. This is called dynamical chiral symmetry breaking [11–13]. The Nambu-Goldstone boson in this system is pion, which is not completely massless due to the current quark masses. The chiral perturbation theory suggests that it is a first order phase transition. It indicates the existence of a critical point at high temperature with some finite chemical potential [14, 15], because the transition at vanishing chemical potential is a crossover. Beam energy scans are performed for the search of a hint of the critical point in nuclear collision experiments. It is also proposed that there could be the second critical point at the lower temperature end of the transition line due to the quark-hadron duality [16].

The phase diagram of low temperature and high density regions is not well known as it is experimentally more challenging to reach, and so far various theories have been proposed. At the high baryon density limit, it is expected from QCD that the light quarks, up, down and strange, form Cooper pairs where three colors and three

[1] It is note-worthy that this is recorded as the highest temperature ever achieved by human beings [10].

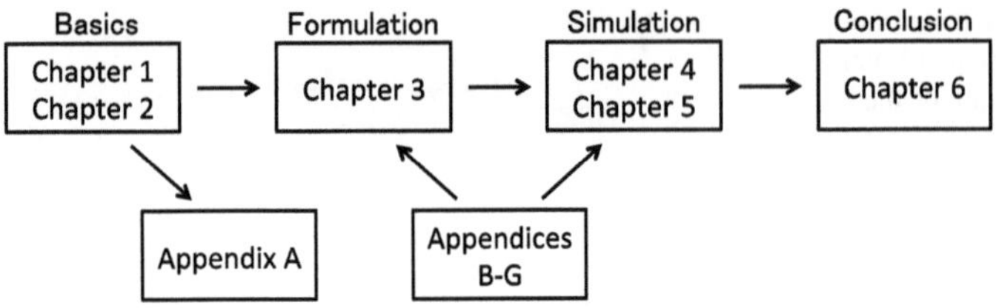

Fig. 1.3 The overall structure of the thesis

flavors are locked to each other. This is known as color-flavor locking, and the system is expected to exhibit color superconductivity [17]. The color-flavor locked phase could be found inside neutron stars. Chiral symmetry is broken in this phase.

1.3 Outline

The thesis is organized as follows. The status of the QGP studies at high-energy heavy ion colliders are summarized in Chap. 2. Past important achievements and future prospects are reviewed with experimental data and theoretical models, along with the meaning of hydrodynamic analyses in the following chapters. Chapter 3 is devoted to the novel formulation of causal relativistic dissipative hydrodynamics. The formalism allows one to describe the systems with conserved net charges such as baryon number and electric charge. This would be suitable for describing the hot QCD medium in collider experiments, unlike the conventional formalism that explicitly assumes elastic binary collisions. In Chap. 4, viscous hydrodynamic model is developed from the hydrodynamic equations derived in Chap. 3 at the vanishing chemical potential limit for the numerical analyses of the effects of secondary interactions on the distribution of the color glass condensate, a pre-collision state, in collider experiments. This sheds light on the issue of excessive charged particle production in a latest result from the Large Hadron Collider. In Chap. 5, relativistic hydrodynamic model with viscosity and dissipation is constructed at finite density. The analyses on the net baryon distributions suggest that more energy is available for the production of the QGP than it is naïvely implied from experimental data. Conclusions and outlook is presented in Chap. 6. The thesis is supplemented with appendices; Appendix A explains a conversion scheme from viscous hydrodynamic picture to particle picture for post-hydrodynamic stages of the collisions. Appendices B-G present thermodynamics and kinetic theory in relativistic systems and useful dissipative hydrodynamic relations. The overall structure of the thesis is summarized in Fig. 1.3. The natural unit $c = \hbar = k_B = 1$ and the Minkowski metric $g^{\mu\nu} = \mathrm{diag}(+, -, -, -)$ is used throughout this chapter.

References

1. Y. Nambu, *A Systematics of Hadrons in Subnuclear Physics, Preludes in Theoretical Physics*, (North Holland Publishing Company, Amsterdam, 1966), p. 133
2. D.J. Gross, F. Wilczek, Asymptotically free Gauge theories. 1. Phys. Rev. D **8**, 3633 (1973)
3. D.J. Gross, F. Wilczek, Asymptotically free Gauge theories. 2. Phys. Rev. D **9**, 980 (1974)
4. S. Borsanyi, G. Endrodi, Z. Fodor, A. Jakovac, S.D. Katz, S. Krieg, C. Ratti, K.K. Szabo, The QCD equation of state with dynamical quarks. JHEP **1011**, 077 (2010)
5. J.M. Maldacena, The large N limit of superconformal field theories and supergravity. Adv. Theor. Math. Phys. **2**, 231 (1998)
6. P. Kovtun, D.T. Son, A.O. Starinets, Viscosity in strongly interacting quantum field theories from black hole physics. Phys. Rev. Lett. **94**, 111601 (2005)
7. K. Fukushima, T. Hatsuda, The phase diagram of dense QCD. Rept. Prog. Phys. **74**, 014001 (2011)
8. G.S. Bali, QCD forces and heavy quark bound states. Phys. Rept. **343**, 1 (2001)
9. K. Yagi, T. Hatsuda, Y. Miake, Quark-Gluon plasma: from big bang to little bang. Camb. Monogr. Part. Phys. Nucl. Phys. Cosmol. **23**, 1 (2005)
10. http://www.guinnessworldrecords.com/records-3000/highest-man-made-temperature/
11. Y. Nambu, G. Jona-Lasinio, Dynamical model of elementary particles based on an analogy with superconductivity. I. Phys. Rev. **122**, 345 (1961)
12. Y. Nambu, G. Jona-Lasinio, Dynamical model of elementary particles based on an analogy with superconductivity. II. Phys. Rev. **124**, 246 (1961)
13. T. Hatsuda, T. Kunihiro, QCD phenomenology based on a chiral effective Lagrangian. Phys. Rept. **247**, 221 (1994)
14. M. Asakawa, K. Yazaki, Chiral restoration at finite density and temperature. Nucl. Phys. A **504**, 668 (1989)
15. A. Barducci, R. Casalbuoni, S. De Curtis, R. Gatto, G. Pettini, Chiral symmetry breaking in Qcd at finite temperature and density. Phys. Lett. B **231**, 463 (1989)
16. T. Hatsuda, M. Tachibana, N. Yamamoto, G. Baym, New critical point induced by the axial anomaly in dense QCD. Phys. Rev. Lett. **97**, 122001 (2006)
17. M.G. Alford, K. Rajagopal, F. Wilczek, Color flavor locking and chiral symmetry breaking in high density QCD. Nucl. Phys. B **537**, 443 (1999)

Chapter 2
High-Energy Heavy Ion Collisions

2.1 Overview

Nuclear collisions have been playing an important role in high-energy nuclear physics as they provide quite unique opportunities to experimentally approach the quark matter under extreme conditions. Prior to the nuclear collisions, observation of the cosmic ray events are the main means of exploring high-energy physics. It is noteworthy that the theoretical studies for the analyses of multi-particle production had built part of the foundation of collider physics today [1]. The heavy ion collisions have the advantage of being able to study the hot medium with larger volume in detail under controlled environments, as beam energy and colliding nuclei can be changed to explore wider regions in the phase diagram of QCD. The collider experiments have been providing great chances and challenges for developing realistic theories of the many-body systems of quarks and gluons including the quark-gluon plasma.

2.1.1 Past Achievements

One of the earliest experiments of heavy ion collisions dates back to Bevalac in Lawrence Berkeley National Laboratory. The heavy ion collisions with more higher energies were carried out in the Alternating Gradient Synchrotron (AGS) at the Brookhaven National Laboratory (BNL) for Au nuclei at $\sqrt{s_{NN}} = 5\,\text{GeV}$ and the Super Proton Synchrotron (SPS) at the European Organization for Nuclear Research (CERN) for Pb nuclei at $\sqrt{s_{NN}} = 17\,\text{GeV}$. Those accelerators were fixed-target experiments. While the energy would not have been sufficient to fully produce the QGP, their results indicated the existence of collective behavior in heavy nuclear collisions. The observation of the J/Ψ suppression, which is possibly due to the color screening, at SPS also suggests the emergence of primordial medium effects.

One of the biggest achievements of the heavy ion experiments is the discovery of the QGP in the Relativistic Heavy Ion Collider (RHIC) [2–5] at BNL which started

A. Monnai, *Relativistic Dissipative Hydrodynamic Description of the Quark-Gluon Plasma*, Springer Theses, DOI: 10.1007/978-4-431-54798-3_2,

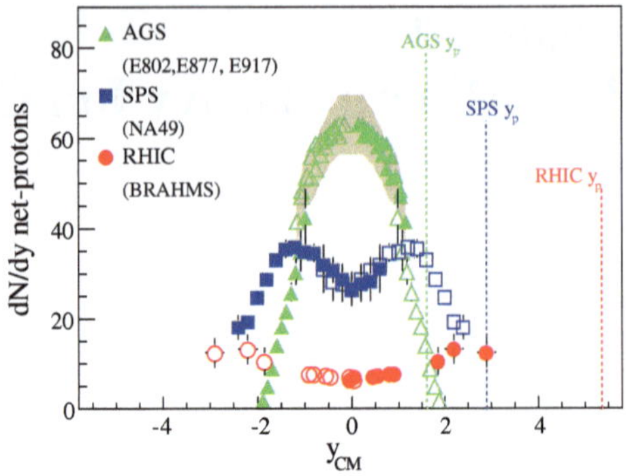

Fig. 2.1 The net proton distribution in the beam direction in momentum space at AGS (E802, E877, E917), SPS (NA49) and RHIC (BRAHMS) experiments for the most central 0–5 % events. (Figure from Ref. [6], American Physical Society.)

in 2000. Designed for collider experiments, the facility has performed Cu–Cu and Au–Au collisions at $\sqrt{s_{NN}} = 130\,\text{GeV}$ and $\sqrt{s_{NN}} = 200\,\text{GeV}$. It is note-worthy that AGS is used as an early-stage accelerator and injector for RHIC. The existence of the hot and dense matter is strongly indicated because sizable medium effects are observed in comparison to the results of proton–proton collisions. One example is jet quenching, where a small bunch of high momentum hadrons called mini-jet gets suppressed when it travels through the bulk medium due to the strong interaction with the medium. The hot medium is expected to be the first human-made QGP since the temperature of the quark-gluon system at RHIC estimated from the photon radiation is above the crossover temperature suggested from the lattice QCD calculations. Estimated energy density of the hot medium is $\sim 5\,\text{GeV/fm}^3$ at RHIC.

The net baryon distribution, which is defined as the distribution of baryons minus that of antibaryons, at AGS, SPS and RHIC is shown in Fig. 2.1 [6] to illustrate how the remnant of shattered colliding nuclei evolves in the heavy ion collisions. The average rapidity of the net baryon number tends to be smaller than the beam rapidity because part of the energy is converted to the production of the hot medium. This phenomenon is known as *baryon stopping*. It is notable that the collisions become increasingly transparent for the nuclear collisions with higher center-of-mass energies. Since the momentum distribution respects the coordinate-space one, this would indicate the creation of a hot medium with finite volume.

Amongst the great achievements at RHIC, it was very surprising that the QGP was found as a nearly-perfect relativistic fluid in the vicinity of the quark-hadron crossover since many had speculated the QGP to be a gas because of the asymptotic freedom of QCD. This is shown by comparing the anisotropy in momentum space [7] against that in coordinate space. The QGP typically has an elliptic shape in the transverse plane for non-central collisions as explained in detail in Sect. 2.2. If the interaction among the constituent quasi-particles is weak, the system is gas-like and the momentum anisotropy would not be produced; on the other hand, if the interaction is strong, i.e. the system is liquid-like, the momentum distribution would reflect the spatial azimuthal anisotropy. The experimental data [8, 9] show that the existence of sizable anisotropy in momentum space and it reaches the hydrodynamic limit

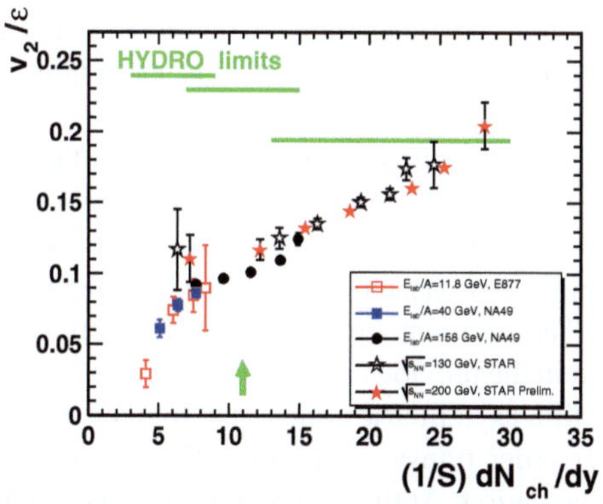

Fig. 2.2 The ratio of azimuthal ellipticity in momentum space v_2 to that in coordinate space ε over the charged particle yield per transverse unit area, which increases as the collision energy does, for AGS (E877), SPS (NA49) and RHIC (STAR) experiments compared with the hydrodynamic expectations. (Figure from Ref. [10], American Physical Society.)

at $\sqrt{s_{NN}} = 200\,\text{GeV}$ collisions (Fig. 2.2 [10]). $(1/S)dN_{\text{ch}}/dy$ in the figure means the yield of charged hadrons per rapidity over the overlapping area of the colliding nuclei, which tends to increase as the energy increases.

The RHIC experiments also allow one to peer into various high-energy phenomena such as gluon saturation in the cold nuclear matter, i.e., a matter at high energies without the effects of hot medium. The evidence of the gluon saturation in the colliding nuclei is considered to be found because the color glass condensate picture (Sect. 2.4.1) gives a good description of both d-Au and Au–Au collisions at RHIC. Note that the hot medium would be produced only in the latter environment.

2.1.2 Current Status and Future Prospects

With the beginning of the heavy ion program at the Large Hadron Collider (LHC) in CERN in 2010, heavy ion physics has entered a new energy regime. LHC has run Pb–Pb collisions at $\sqrt{s_{NN}} = 2.76\,\text{TeV}$, and its first data started to provide valuable information for the quark matter on the high temperature frontier. The currently-planned goal for the collision energy is $\sqrt{s_{NN}} = 5.5\,\text{TeV}$. It is believed that the properties of the hot medium does not fundamentally change from RHIC to LHC [11], though several intriguing anomalies are reported in particle production [12]. The analyses of azimuthal anisotropy show that the medium still behaves as a fluid with small viscosity, which is important information since it has been naïvely expected that the QGP becomes slightly more weakly-coupled with increasing energy due to the QCD asymptotic freedom. On the other hand, since it is the first experiment to explore the energy dependence of the QGP, it had been possible that the system became too strongly-coupled even for ideal hydrodynamic description, in which case one would need a different mechanism for explaining the data. The energy density at LHC is estimated to be ~15 GeV/fm^3, which is about three times larger than that at RHIC.

RHIC has started to explore the system dependence of the hot matter. The facility recently performed the U–U collision at $\sqrt{s_{NN}} = 193$ MeV and the Au–Cu collision at $\sqrt{s_{NN}} = 200$ MeV [13]. The former is the collision of nuclei with prolate deformation, and thus tip-to-tip collisions and body-to-body collisions are expected to show different properties. The uranium tip-to-tip collisions produce more dense matter than what is produced in the previous Au–Au collisions by ~30 % in the energy density. The latter is the first asymmetric collision at high energies. It provides intrinsic triangular and higher order transverse anisotropies in momentum space that does not originate from fluctuation. At the most central events, Cu nucleus is going to be completely buried in Au nucleus, producing the QGP in a corona of nucleons.

The search for the QCD critical point on the phase diagram is also a very important topic. One can access a wider region on the T-μ_B plane by changing the beam energy, because the net baryon chemical potential becomes higher as the temperature becomes lower and vise versa. The beam energy scans (BES) are performed at RHIC at $\sqrt{s_{NN}} = 62.4, 39, 27, 19.6, 11.5, 7.7$ and 5.5 GeV [14]. Although the critical point is yet to be found, the momentum anisotropy is found to persist until around 19.6 MeV [15]. BES experiments are also performed at SPS for light nuclei at $\sqrt{s_{NN}} = 4.9 - 17.3$ MeV, and planned at Nuclotron-based Ion Collider fAcility (NICA) in Joint Institute for Nuclear Research with Au nuclei at $\sqrt{s_{NN}} = 3.9 - 11$ MeV and at Facility for Antiprotons and Ion Research (FAIR) in GSI Helmholtzzentrum für Schwerionenforschung with Au nuclei for $\sqrt{s_{NN}} = 4.9 - 17.3$ MeV. Japan Proton Accelerator Research Complex (J-PARC) also has a plan for low energy collisions.

2.2 Geometrical Setup

The two nuclei collide nearly at the speed of light in high-energy heavy ion collisions. Thus they are squeezed in the direction of beam axis due to Lorentz contraction in the laboratory frame. At RHIC energy $\sqrt{s_{NN}} = 200$ GeV, Lorentz dilation factor is $\gamma \sim 100$ for a projectile nuclei, which means the nucleus of diameter ~14 fm is reduced to ~0.1 fm. Likewise at LHC energy $\sqrt{s_{NN}} = 2.76$ TeV, $\gamma \sim 1500$ and the nucleus is squeezed to ~0.01 fm. The hot medium would be produced in the overlapping area between the two passing nuclei. The collision axis is conventionally chosen as z-axis, and often referred to as the longitudinal direction as opposed to the transverse plane, which is perpendicular to the collision axis. Collisions can be non-central, and the resulting geometry of a hot medium is almond-shaped in the transverse directions. The nucleons which collide are called participants, and those which do not are called spectators. The schematic pictures of the collision geometry of symmetric nuclei are shown in Fig. 2.3.

It is more convenient to introduce the relativistic τ-η_s coordinate system to describe the heavy ion systems, where

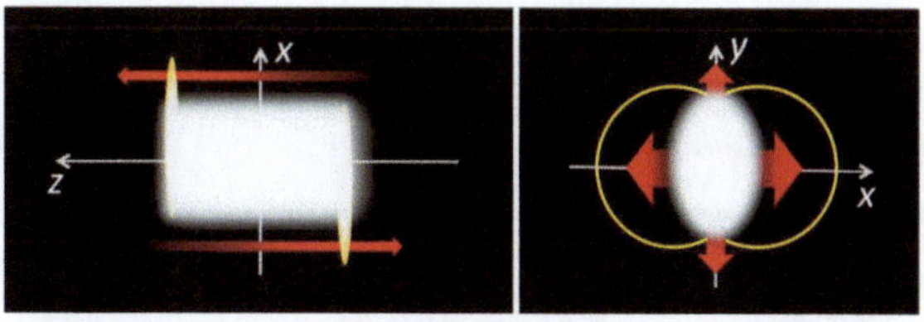

Fig. 2.3 Schematic picture of the geometry of non-central high-energy heavy ion collisions for (*left*) the longitudinal relativistic expansion and (*right*) the transverse expansion

$$\tau = \sqrt{t^2 - z^2}, \tag{2.1}$$

$$\eta_s = \frac{1}{2} \ln \frac{t+z}{t-z}, \tag{2.2}$$

are the proper time and the space-time rapidity. The space-time rapidity is a dimensionless variable that can be interpreted as a hyperbolic angle. They satisfy the relations $t = \tau \cosh \eta_s$ and $z = \tau \sinh \eta_s$. $\eta_s = 0$ corresponds to the t axis and $\eta_s = \pm\infty$ the light cone. Similarly, one defines the transverse mass m_T and the rapidity y in momentum space as

$$m_T = \sqrt{E^2 - p_z^2}, \tag{2.3}$$

$$y = \frac{1}{2} \ln \frac{E + p_z}{E - p_z}. \tag{2.4}$$

In collider physics, the transverse momentum $p_T = \sqrt{m_T^2 - m^2}$ and the pseudo-rapidity $\eta_p = \frac{1}{2} \ln [(|p| + p_z)(|p| - p_z)]$ are useful variables because they are independent of mass and thus of particle species. At relativistic energies, they are also fairly close to the transverse mass and the rapidity, respectively, and become identical in relativistic massless limit.

The polar coordinate system is often employed in analyses of the transverse dynamics. The angle in the configuration space is denoted as ϕ and that in the momentum space as ϕ_p. They are related to the variables in Cartesian coordinates as $(x, y) = (r \cos \phi, r \sin \phi)$ and $(p_x, p_y) = (p_T \cos \phi_p, p_T \sin \phi_p)$.

The non-centrality of collisions is characterized by the impact parameter b, which is defined as the distance between the centers of the two colliding nuclei on the transverse plane. *Centrality* is defined by groups of events per the number of participants because the collision would have more participant nucleons for more central collisions. The groups are ordered from the most central events, e.g., 0–20 % centrality means that the most central collisions are selected up to 20 % of the total events, 20–40 % centrality the next 20 % events and likewise. The variable is preferred in collider experiments as the impact parameter is not a direct observable. The number of participants still reflects the centrality in asymmetric collisions where the smaller nuclei is buried in the larger one because of the effects of fluctuating geometry.

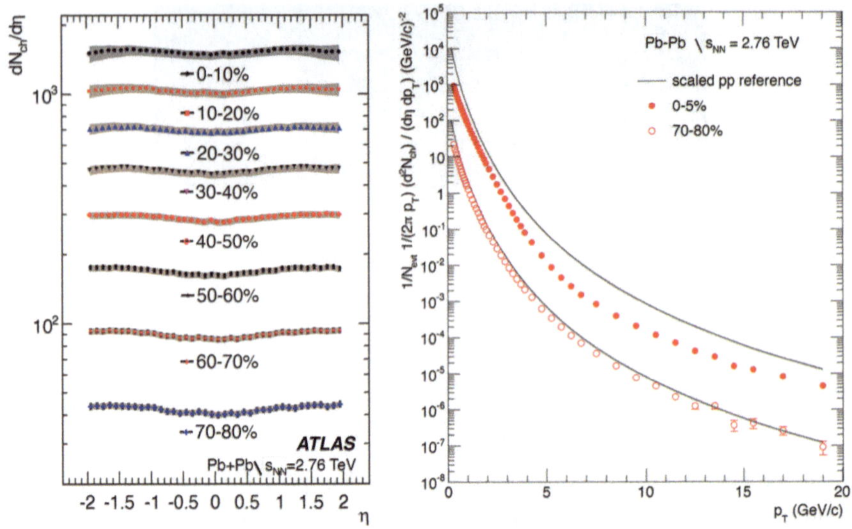

Fig. 2.4 (*Left*) The pseudo-rapidity distributions of the charged particles for different centralities in Pb–Pb collisions at 2.76 TeV by ATLAS Collaboration. (Figure from Ref. [16], Elsevier.) (*Right*) The p_T spectra of the charged particles for central and peripheral collisions in the same collisions by ALICE Collaboration. (Figure from Ref. [21], Elsevier.)

2.3 Experimental Observables

2.3.1 Hadronic Particle Spectra

Hadronic particle yield is an essential observable to study high-energy heavy ion collisions, since hadrons constitute the bulk part of the produced medium, i.e., most of the initial energy is carried by hadrons. Various insights can be obtained by analyzing its dependences on transverse momentum, rapidity, centrality, beam energy and particle species. Due to the strong interaction in the medium, the particle spectra is considered to contain the information at the latest stage of heavy ion collisions. Thus detailed profiles of QGP time evolution is integrated out from the observable.

The rapidity distribution dN/dy, or the pseudo-rapidity distribution $dN/d\eta_p$, is a basic observable to quantify particle production in the system (Fig. 2.4, left [16]). The charged particle multiplicities at mid-rapidity are $dN_{\mathrm{ch}}/d\eta_p \sim 650$ at RHIC ($\sqrt{s_{NN}} = 200\,\mathrm{GeV}$) and $dN_{\mathrm{ch}}/d\eta_p \sim 1600$ at LHC ($\sqrt{s_{NN}} = 2.76\,\mathrm{GeV}$) in the most central events [11, 16–20]. It is note-worthy that the multiplicity of charged particles at LHC was larger than most of the theoretical predictions made prior to the experiment. A possible solution to the discrepancy will be presented in Chap. 4. The rapidity distribution of the net baryon number originates only from the colliding nuclei and thus is sensitive to kinetic energy loss of the nucleons which becomes available for the production of the medium as seen in Sect. 2.1.1. This would be revisited later in Chap. 5.

The transverse momentum distribution is very informative (Fig. 2.4, right [21]). The Fourier harmonics in the series expansion of p_T spectra is an essential observable since the hot medium is almond-shaped in non-central collisions [7, 22]. The particle spectrum is expressed as

$$\frac{dN}{d\phi_p p_T dp_T dy} = \frac{1}{2\pi}\frac{dN}{p_T dp_T dy}\left[1 + 2\sum_n v_n(p_T, y)\cos(n\phi_p - n\Psi)\right], \quad (2.5)$$

with Fourier harmonics $v_n(p_T, y)$ and the reaction plane Ψ. This leads to

$$v_n(p_T, y) = \int d\phi_p \cos(n\phi_p - n\Psi)\frac{dN}{d\phi_p p_T dp_T dy}\bigg/\int d\phi_p \frac{dN}{d\phi_p p_T dp_T dy}. \quad (2.6)$$

The space-time azimuthal anisotropy in the hot medium suggests that the second order harmonics v_2, *elliptic flow*, would be larger than other harmonics for non-central collisions in heavy ion collisions. If one takes the reaction plane as the minor axis of the ellipsoid based on the average over many events, the odd-order harmonics vanishes because of the symmetry around the collision axis.[1] The large v_2 observed in the high-energy heavy ion collisions is well quantified by hydrodynamic models, supporting the fact that the QGP is a strongly-coupled medium. It is known to be roughly proportional to the spatial anisotropy ε_2 at RHIC and LHC. The elliptic flow is known to be very sensitive to viscosity in the hydrodynamic phase, as deviation from equilibrium would lead to a less strongly-coupled medium. The experimental data typically shows agreement with the hydrodynamic calculations up to $p_T \sim 1$ GeV if there is no viscosity, and $p_T \sim 3$ GeV if shear viscosity is introduced. The flow harmonics can be used to experimentally constrain the transport coefficients in the medium. It should be noted that the applicability of hydrodynamics naturally breaks down for high-p_T regions as quasi-particles do not reach local equilibrium and v_2 decreases as the deviation from the fluid picture becomes larger.

Higher flow harmonics should remain finite in event-by-event analyses since initial conditions have geometrical fluctuations that break the symmetry [23, 24]. This provides new sets of observables such as triangular flow v_3 for the analyses of a hot medium (Fig. 2.5). Here the variable would be defined as

$$v_n^{\mathrm{ebe}}(p_T, y) = \int d\phi_p \cos(n\phi_p - n\Psi_n)\frac{dN}{d\phi_p p_T dp_T dy}\bigg/\int d\phi_p \frac{dN}{d\phi_p p_T dp_T dy}, \quad (2.7)$$

where Ψ_n is chosen for each event and harmonics. The analyses of the observed higher flow harmonics [25, 26] are important in precision physics.

[1] Directed flow v_1 is the exception because in non-central collisions, rapidity-dependent flow asymmetry is created as the expansion gets tiled from the collision axis. It is an odd function of rapidity, and should be distinguished from the one from transverse geometrical fluctuation, which would be an even function of rapidity.

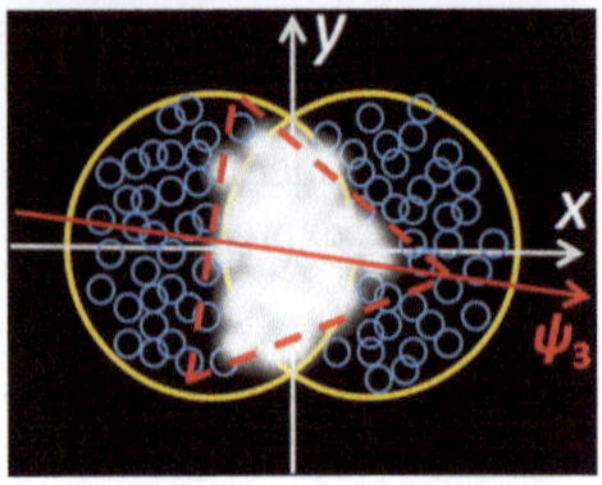

Fig. 2.5 A schematic picture of the triangular anisotropy and the event plane Ψ_3 originating from the fluctuation in an initial condition

The particle spectrum in a high-p_T region ($p_T \geq 5\,\text{GeV}$) shows non-hydrodynamic behavior. At peripheral collisions the spectrum exhibits simple power law behavior which is expected from perturbative QCD in Fig. 2.4 (right). At central collisions, it is clearly less than the scaled reference because the medium effects become more prominent. The deviation is quantified by the nuclear modification factor R_{AA} as

$$R_{\text{AA}}(p_T, y) = \frac{dN_{\text{AA}}}{p_T dp_T dy} \Big/ \langle N_{\text{coll}} \rangle \frac{dN_{pp}}{p_T dp_T dy}, \tag{2.8}$$

where $\langle N_{\text{coll}} \rangle$ is the average number of binary collisions. AA denotes a nucleus–nucleus collision and pp a proton–proton reference. R_{AA} is mostly smaller than 1 for hadrons in central collisions. At low energy collisions, it can become larger than unity due to the multiple parton scattering called Cronin effect. Experimental data shows that the Cronin effect is not large enough to counter the suppression in high-energy nucleus–nucleus collisions.

2.3.2 *Jets and Quarkonia*

Jet quenching is a very strong evidence of the presence of a hot and dense medium, the concept of which is originally proposed in the context of hadron–hadron collisions [27]. In nucleus–nucleus collisions, a pair of streaks of partons at high momentum called (mini)-*jets* are produced. They are not thermalized and considered to be dominant in hadronic particle spectra at high-p_T. The triggered event is called the near-side jet, and the other side the away-side jet. For proton–proton binary collisions and deuteron-gold collisions, the two streaks are clearly observed [28]. On the other hand, the away-side jet is absorbed, if not completely, in nucleus–nucleus collisions [29]. The results are summarized in Fig. 2.6 (left) [28]. This phenomenon is called jet quenching, and is considered to indicate the creation of a chromodynamically interacting hot matter only in the Au–Au collisions. The magnitude of parton energy loss is expected to be sensitive to the density properties of a medium. The away-side jet is reduced to low momentum partons spread in wide angle and the rest of energy is expected to go back to the medium itself.

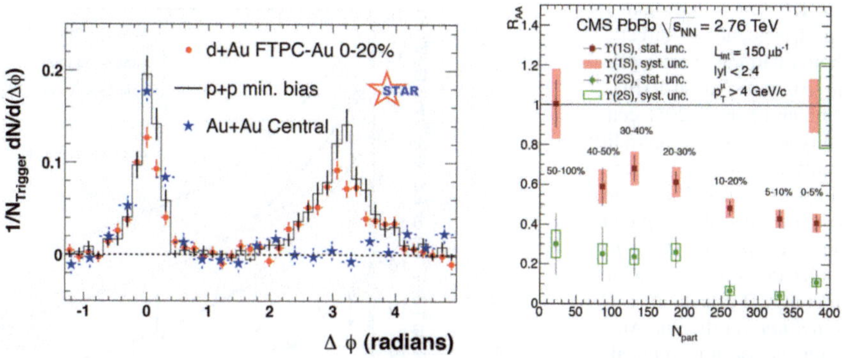

Fig. 2.6 (*Left*) The correlation of the near-side and the away-side jets ($4 < p_T < 6$ GeV) are clearly visible for p-p and d-Au collisions, whereas it is lost for Au–Au collisions at $\sqrt{s_{NN}} = 200$ GeV. (Figure from Ref. [28], American Physical Society.) (*Right*) R_{AA} for $\Upsilon(1S)$ and $\Upsilon(2S)$ as a function of the number of participants at Pb–Pb collisions from CMS Collaboration at LHC. (Figure from Ref. [31], American Physical Society.)

Quarkonia, bound states of a heavy quark and its antiquark, are expected to be able to survive in the QGP as the lattice QCD calculations indicate that the binding energy for the ground state of charmonium $J/\Psi(1S)$ is ~ 0.6 GeV and that of bottomonium $\Upsilon(1S)$ is ~ 1.2 GeV, which are higher than the QCD scale $\Lambda_{\mathrm{QCD}} \sim 0.2$ GeV. On the other hand, the quarkonia melt if the QGP temperature is sufficiently high because of the color screening effect. The higher excited states are less tightly bound and thus easier to melt as the temperature becomes higher. The phenomenon is called sequential melting. This suggests that quarkonia can be used as a thermometer of the produced medium. The suppression of the charmonium production in medium is called J/Ψ suppression [30]. The experimental data show a clear evidence of the suppression of J/Ψ at SPS, RHIC and LHC. It is note-worthy that J/Ψ regeneration process in later stages may be important for quantitative analyses in higher energy collisions. The LHC also observes Υ suppression where the excited state is more suppressed than the ground state (Fig. 2.6, right) [31]. It is interesting to note that the J/Ψ suppression in the Au–Cu collisions [13] shows stronger suppression for the Cu-going side whereas it remains the same for the Au-going side compared with that in the Au–Au collisions.

2.3.3 Electroweak Probes

The experimental data indicate that the quark-gluon plasma is a very opaque medium with respect to the color charge. On the other hand, it is suggested to be reasonably transparent in terms of electromagnetic interactions. This motivates one to use electromagnetically charged particles with no color as probes for the hot matter since

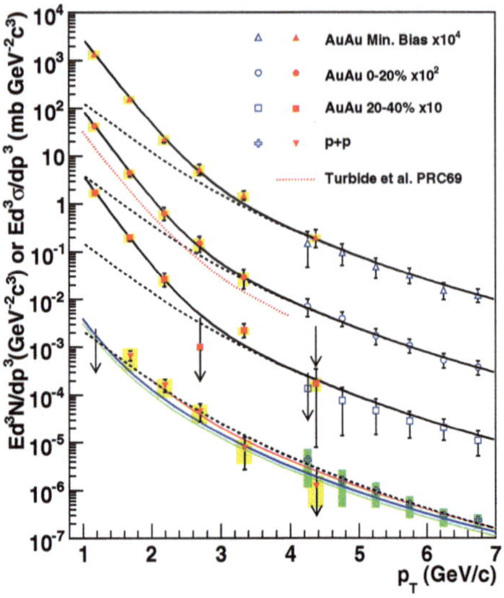

Fig. 2.7 Transverse momentum dependence of direct photon particle spectra of Au–Au collisions at RHIC. Minimum bias, 0–20 % centrality and 20–40 % centrality events are shown together with *p*-*p* results. The *dashed lines* are the scaled *p*-*p* data. The three *lines* on the *p*-*p* results are of next-to-leading order pQCD calculations. The *dotted line* on the Au–Au results is also a theoretical calculation. (Figure from Ref. [32], American Physical Society.)

they should retain the information of the medium at the time of their creation unlike the colored observables.

Photons play an important role in estimating the temperature of the hot medium. There are multiple sources of photons in heavy ion collisions. The photons created at the time of a collision are called *prompt photons*. This is a heavy-ion analogy to the cosmic microwave background in the early universe. The photons emitted from the QGP as an analogy to the black-body radiation is called *thermal photons*, which is relevant to the estimation of the medium temperature. Finally, photons are also produced when hadrons decay in later stages; this is called *decay photons*. Decay photons are the majority (~90 %) of the inclusive photons for $1 < p_T < 3\,\text{GeV}$ where the thermal photons become important. The prompt and thermal photons together are called direct photons. Direct photon spectra for Au–Au and *p*-*p* collisions are shown along with next-to-leading order pQCD calculations in Fig. 2.7 [32]. One can find clear enhancement of photon yields for the Au–Au data. The comparison with the fit by an exponential function of the slope parameter—or the effective temperature—yields $221 \pm 38\,\text{MeV}$ for the medium that emitted thermal photons in the central 0–20 % collisions. Since this is the average over time evolution, the initial temperature T_{init} should be larger. Hydrodynamic model calculations suggest $T_{\text{init}} \sim 300$–$600\,\text{MeV}$ for the thermalization time $\tau_0 \sim 0.6$–$0.15\,\text{fm}/c$. This is well beyond the crossover temperature $T_c \sim 170\,\text{MeV}$, which strongly suggests the hot medium is the quark-gluon plasma rather than a hadronic matter at RHIC. The enhancement is not observed for *d*-Au collisions. Latest experiments at LHC indicate a slightly higher slope parameter 304 MeV [33] for the thermal photon production on time average.

One can also consider the elliptic flow of photons in analogy to the hadronic case. The elliptic flow v_2 of direct photons reflects the thermalization time of the QGP because prompt photons do not have v_2 since the momentum anisotropy has not developed yet at time zero. Theoretical estimation suggests that if the thermalization is early, v_2 of thermal photons become small. This seems rather counterintuitive but could be understood that the energy density decreases faster in hydrodynamic evolution and the elliptic flow cannot fully develop for too early thermalization. v_2 of direct photons can be as large as that of hadrons for late thermalization times. The latest experimental data [34] show that there is a serious underestimation by the theoretical estimations of hydrodynamics, making it one of the important clues to uncover the yet-unknown features of the hot medium.

The production of W and Z bosons in Pb–Pb collisions at LHC, on the other hand, follows that of p-p collisions scaled by the number of participants very well [35, 36]. This suggests that they are created at the time of the collision and do not interact with the QCD medium. The same trend can be found for large momentum photons at RHIC and LHC.

2.4 Theoretical Modeling

The QCD system in high-energy heavy ion collisions goes through various stages from nuclei to the quark-gluon plasma to hadrons. Since the system is suggested to be strongly coupled, one would not be able to naïvely employ perturbative QCD for estimating the evolution of the bulk medium. The dynamical first-principle simulation of a non-perturbative QCD medium at finite temperature and chemical potential is currently beyond the capacity of any theoretical method developed so far. The fact that collective behavior exists in the hot medium, on the other hand, allows one to build an effective model of the hot matter. It is sometimes referred to as a hydrodynamic model, but hydrodynamic picture is applicable for the intermediate stage in heavy ion collisions around $\tau \sim \mathcal{O}(1)$–$\mathcal{O}(10)$ fm/c. A "standard model" of high-energy nucleus–nucleus collisions consists of the color glass condensate, glasma, relativistic hydrodynamics and hadronic gas (Fig. 2.8) which will be explained in order in the following sections.

2.4.1 Color Glass Condensate

The colliding nuclei is accelerated to near-light speed in high-energy nuclear collisions. At those energies, the life-time of gluons emitted from the valence quarks or other gluons become long enough to allow additional emissions of soft gluons from themselves. The cascading process keeps increasing the number density of gluons until gluons become saturated as recombination of gluons becomes non-negligible. This state of matter is called color glass condensate (CGC) [37–49]. The nonlinear behavior is described in Balitsky–Kovchegov equation [43, 44]. The gluon occupation number is typically of order $\sim 1/\alpha_s$. The gluon emission can be characterized

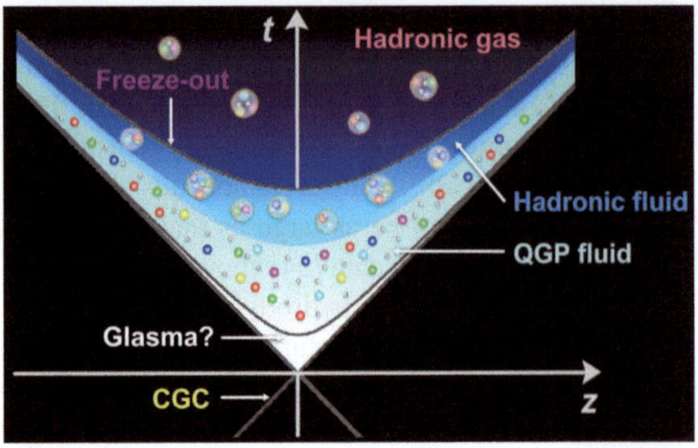

Fig. 2.8 A "standard model" of the space-time evolution of a hot medium in high-energy heavy ion collisions. The hyperbolic geometry is due to the Lorentz dilation of relativistic longitudinal expansion

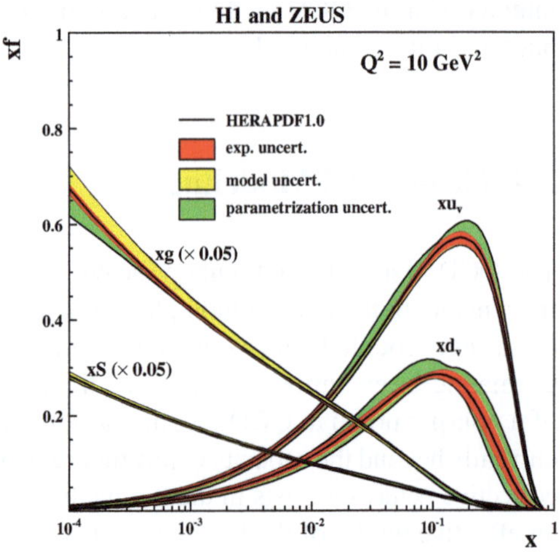

Fig. 2.9 The distributions of valence quarks $xu_v(x)$ and $xd_v(x)$, sea quarks $xS(x)$ and gluons $xg(x)$ as functions of x for the proton from H1 and ZEUS Collaborations at HERA. (Figure from Ref. [50], Springer Science+Business Media.)

by momentum fraction x of incident particles. It gets smaller as more partons are produced at higher energies. Figure 2.9 shows the distribution functions of valence up and down quarks, sea quarks and gluons observed at Hadron Electron Ring Accelerator (HERA) experiments for protons [50]. The valence quarks are important for large x and their distributions peak around $x \sim 1/3$. On the other hand, the gluons become dominant against the valence quarks for small x. The gluon saturation is suggested to be realized in high-energy heavy ion collisions since $x \sim 10^{-2}$ at RHIC and $x \sim 5 \times 10^{-4}$ at LHC. Saturation scale $Q_s(x)$ is the typical momentum of a gluon when the gluons starts to overlap in the transverse plane. It is parametrized as $Q_s^2(x, A) \sim A^{1/3}x^{-\lambda}$ where A is the atomic number of the nucleus. λ is a parameter that is experimentally suggested to be in the range 0.2–0.3 [51, 52].

The color glass picture has been successful in reproducing the centrality dependence of charged particle yields for the deuteron/proton–nucleus and nucleus–nucleus collisions assuming the number of hadrons much that of gluons at RHIC, and various studies have been made for LHC as well [53–64]. Its relation to the hydrodynamic space-time evolution for Au–Au collisions at RHIC and Pb–Pb collisions at LHC will be investigated later in Chap. 4.

2.4.2 Early Thermalization

It is implied from the experiments that the system locally thermalizes in a surprisingly short time ($\sim$0.6–1 fm/c) after the collisions. The precise mechanism of this early thermalization is not fully known. If one solves Yang–Mills equation [65] with the CGC initial condition, color tube structures appear between the collided nuclei, which is known as *glasma* [66]. A typical transverse size of the tube is $\sim 1/Q_s$. The word is a mixture of the color *glass* condensate and the quark-gluon *plasma*. Although it is expected to provide a bridging picture between the two models towards thermalization, the glasma currently has the problem that the longitudinal pressure stays negative during its time evolution and never reaches the hydrodynamic limit where the pressure is almost isotropic in all directions. One possible solution to the problem is instability in the glasma [67–71]. The non-boost invariant picture may play an important role in the longitudinal thermalization [72]. It is also reported that isotropization may be encouraged by the Bose–Einstein condensate that might be realized in the medium [73].

2.4.3 Hydrodynamic Stage

The correlation between the azimuthal anisotropy in coordinate space and that in momentum space at high energies suggests that the quark-gluon plasma would need to be described by relativistic hydrodynamic picture for a certain period of its lifetime. It is note-worthy that the description of relativistic hydrodynamics apply not only to the quark-gluon plasma but also to the hadronic matter near the pseudo-critical temperature as well, making it a simple and effective method to treat the time-evolution in the crossover regime. Historically speaking, the relativistic hydrodynamic study of hadron collisions started with the Landau model [74] where the hot medium created in the middle of the two shattered nuclei is speculated have no collective velocity before it starts to expand into vacuum. It should be noted the picture was originally proposed for hadron–hadron collisions. The full nuclear stopping picture later turned out to be rather unrealistic in the early heavy ion experiments since the stopping power is finite, and the hot matter was found to be described better by the scaling expansion in the longitudinal direction. This is called the Bjorken model [75], and to the present day the boost-invariant ansatz can be found in many hydrodynamic and non-hydrodynamic models of relativistic heavy ion collisions.

The fluidity of a hot medium was first observed in analyses of the Au–Au collisions at RHIC. The first relativistic ideal hydrodynamic analyses were performed by Kolb et al. [76–78] (Fig. 2.10, left), followed by Teaney et al. [79, 80]. They

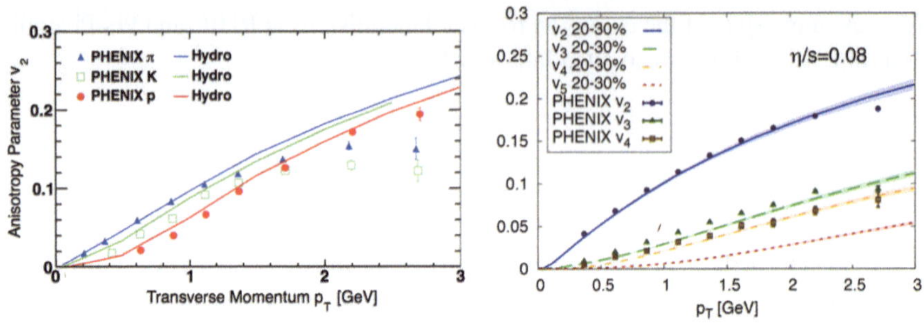

Fig. 2.10 (*Left*) The first ideal hydrodynamic calculations for v_2 of p_T with particle identification (π, K, p) by Huovinen et al. (Figure from Ref. [77], Elsevier.) (*Right*) State-of-art viscous hydrodynamic calculations with fluctuation for v_2, v_3, v_4 and v_5 of p_T for 20–30 % centrality with the shear viscosity to entropy ratio $\eta/s = 0.08$ by B. Schenke et al. (Figure from Ref. [134], American Physical Society.)

have shown that the elliptic flow is quantitatively well-explained by hydrodynamic models, which indicates the existence of the strongly-coupled QGP in high-energy heavy ion collisions. An ideal hydrodynamic model that includes non-boost invariant longitudinal expansion in addition to the transverse one was introduced by Hirano [81, 82]. The equation of state is assumed to be of first order phase transition for those calculations. Various hybrid approaches of ideal hydrodynamic and hadronic transport models are also investigated [79, 80, 83–89] along with improvements in initial conditions and the equation of state by introduction of first principle-based results.

As mentioned in Sect. 2.3.1, the experimentally observed elliptic flow v_2 is described well by ideal hydrodynamic models up to $p_T \sim 1$ GeV, and the picture breaks down for mid-high p_T because those components would not be fully-thermalized. The agreement with the experimental data becomes better by introducing viscosity, which would embody deviation from equilibrium in hydrodynamic systems. Viscous flow analyses have been performed in boost-invariant geometry with shear viscosity [90–112] and also with shear and bulk viscosities [113–118]. Elliptic flow results with viscous corrections typically show good agreements with the data up to $p_T \sim 3$ GeV. Systematic studies on parameter dependences are also performed [119, 120]. The higher-order harmonics are explained in event-by-event hydrodynamic models well, further strengthening the understanding that the QGP is a relativistic fluid [121–125]. Hadronic transport picture for the later stages of the collision is taken into account in Refs. [126–128].

All the viscous hydrodynamic calculations mentioned above assumes boost-invariant expansion and vanishing baryon chemical potential possibly due to the difficulty in viscous formalism of relativistic expansion. The importance of the longitudinal expansion is being recognized at RHIC and LHC with shear viscosity [129], with shear and bulk viscosities with a CGC initial condition [130] and with viscosities and baryon diffusion at finite density [131]. Figure 2.10 (right) shows state-of-art

(3+1)-dimensional viscous hydrodynamic simulation results by Schenke et al. [132–134], which include shear viscosity. Their work is also followed by shear and bulk viscous calculations by Bozek [135–137] and by yet-another shear viscous simulation by Vredevoogd and Pratt [138].

Recent trends in hydrodynamic studies include analyses with temperature dependent transport coefficients [139–141], which would be important to go beyond the conventional conjecture that the shear viscosity is proportional to the entropy density. Highly-anisotropic hydrodynamic models are venturing frameworks that assume only transverse thermalization [142–145]. This is motivated by the fact that the glasma picture yields the negative longitudinal pressure. The importance of hydrodynamic fluctuation coming from fluctuation–dissipation theorem was also investigated [146]. The detailed analyses on flow harmonics [147, 148] and the event plane correlations [149] have been performed as well.

There are a couple of choices for the hydrodynamic initial condition due to the fact that ambiguities still remain in the early thermalization stage. Monte–Carlo Glauber model is a phenomenological approach, which is constructed by geometrically placing wounded nucleons according to the nuclear density function. Monte–Carlo Kharzeev–Levin–Nardi (MC-KLN) model is based on CGC with k_T-factorization formula [150, 151]. The gluon energy or entropy density distribution is matched with that at the beginning of hydrodynamic evolution. Monte–Carlo running coupling Balitsky–Kovchegov (MCrcBK) is an advanced CGC based method which takes into account the effect of QCD running coupling [62]. IP-Glasma is the combination of impact parameter dependent saturation model (IP-Sat) and the glasma picture [152, 153].

2.4.4 Freeze-Out and Hadronic Transport

The quark-gluon system cools down with time evolution to the point where hydrodynamic description is no longer applicable due to the break-down of the strongly coupled picture. Freeze-out is a concept that is conventionally employed for describing the transition from the liquid picture to the gas one. The boundary between the two pictures form a 3-dimensional hypersurface Σ in Minkowski space-time. The total number of type-i particles N_i on the freeze-out hypersurface in relativistic kinetic theory is given as

$$N_i = \int_\Sigma d\sigma_\mu N_i^\mu = \int_\Sigma d\sigma_\mu \int \frac{g_i dp_i^3}{(2\pi)^3 E_i} p_i^\mu f_i, \tag{2.9}$$

where g_i is the degeneracy of the particle and $\sigma_\mu(x)$ is the local freeze-out hypersurface element which is perpendicular to Σ. The difference from a simple Gauss's law is that particles can be dynamically created and annihilated in the medium. The above equation gives the particle number that is frozen out at the hypersurface. Then its differential form gives the Cooper–Frye formula [154],

$$E_i \frac{dN_i}{d^3p} = \frac{dN_i}{d^2p_T dy} = \frac{g_i}{(2\pi)^3} \int_\Sigma p_i^\mu d\sigma_\mu f_i, \tag{2.10}$$

which gives particle spectra in phase space. The freeze-out hypersurface is usually defined by the freeze-out temperature $T_f \sim 0.13$–$0.16\,\text{GeV}$ but one may choose some other criteria such as the magnitude of flow derivatives [112] which would be sensitive to hydrodynamic applicability. It is note-worthy that when the system is completely in local equilibrium, input from a hydrodynamic model is the flow, the temperature and the chemical potential that appears in the distribution f_i, and the energy density, the pressure and the net charge density are not used. The distribution is subject to viscous corrections, which can add extra modifications to particle spectra at freeze-out [112, 117, 155–157]. Estimations of effects of shear and bulk viscous corrections on the distribution function will be discussed in Appendix A.

The hadronic matter after freeze-out is considered to further evolve in a weakly-coupled picture. Hadronic transport models are originally used for describing a hot matter at the low-energy nuclear collisions at AGS and SPS which does not enter a fully-developed hydrodynamic phase. The heavy baryons and mesons created at freeze-out decay during the hadronic evolution, which is called hadronic cascade. Notable transport models include, but are not limited to, Jet AA Microscopic transport model (JAM) [158] and Ultra-relativistic Quantum Molecular Dynamics (UrQMD) [159, 160]. It is note worthy UrQMD also serves initial conditions for hydrodynamic simulations, which exhibit agreement with experimental data [87–89]. The hadronic transport is considered to be essential in understanding identified particle spectra from the collider experiments quantitatively.

References

1. E. Fermi, High energy nuclear events. Prog. Theor. Phys. **5**, 570 (1950)
2. K. Adcox et al., [PHENIX Collaboration], Formation of dense partonic matter in relativistic nucleus-nucleus collisions at RHIC: experimental evaluation by the PHENIX collaboration. Nucl. Phys. A **757**, 184 (2005)
3. J. Adams et al., [STAR Collaboration], Experimental and theoretical challenges in the search for the quark gluon plasma: the STAR Collaboration's critical assessment of the evidence from RHIC collisions. Nucl. Phys. A **757**, 102 (2005)
4. B.B. Back et al., [PHOBOS Collaboration], The PHOBOS perspective on discoveries at RHIC. Nucl. Phys. A **757**, 28 (2005)
5. I. Arsene et al., [BRAHMS Collaboration], Quark gluon plasma and color glass condensate at RHIC? The Perspective from the BRAHMS experiment. Nucl. Phys. A **757**, 1 (2005)
6. I.G. Bearden et al., [BRAHMS Collaboration], Nuclear stopping in Au + Au collisions at $\sqrt{s_{NN}} = 200$ GeV. Phys. Rev. Lett. **93**, 102301 (2004)
7. J.-Y. Ollitrault, Anisotropy as a signature of transverse collective flow. Phys. Rev. D **46**, 229 (1992)
8. S.S. Adler et al., [PHENIX Collaboration], Elliptic flow of identified hadrons in Au+Au collisions at $\sqrt{s_{NN}} = 200$ GeV. Phys. Rev. Lett. **91**, 182301 (2003)
9. J. Adams et al., [STAR Collaboration], Particle type dependence of azimuthal anisotropy and nuclear modification of particle production in Au + Au collisions at $\sqrt{s_{NN}} = 200$ GeV. Phys. Rev. Lett. **92**, 052302 (2004)

10. C. Alt et al., [NA49 Collaboration], Directed and elliptic flow of charged pions and protons in Pb + Pb collisions at 40A GeV and 158A GeV. Phys. Rev. C **68**, 034903 (2003)
11. K. Aamodt et al., [The ALICE Collaboration], Elliptic flow of charged particles in Pb-Pb collisions at 2.76 TeV. Phys. Rev. Lett. **105**, 252302 (2010)
12. K. Aamodt et al., [The ALICE Collaboration], Charged-particle multiplicity density at mid-rapidity in central Pb-Pb collisions at $\sqrt{s_{NN}}$ = 2.76 TeV. Phys. Rev. Lett. **105**, 252301 (2010)
13. T. Sakaguchi [PHENIX Collaboration], PHENIX highlights. arXiv:1211.0324 [nucl-ex] (2012)
14. M. M. Aggarwal et al. [STAR Collaboration], Anexperimental exploration of the QCD phase diagram: the search for the critical point and the onset of de-confinement. arXiv:1007.2613 [nucl-ex] (2010)
15. L. Adamczyk et al., [STAR Collaboration], Inclusive charged hadron elliptic flow in Au + Au collisions at $\sqrt{s_{NN}}$ = 7.7 - 39 GeV. Phys. Rev. C **86**, 054908 (2012)
16. G. Aad et al., [ATLAS Collaboration], Measurement of the centrality dependence of the charged particle pseudorapidity distribution in lead-lead collisions at $\sqrt{s_{NN}}$=2.76 TeV with the ATLAS detector. Phys. Lett. B **710**, 363 (2012)
17. I.G. Bearden et al., [BRAHMS Collaboration], Pseudorapidity distributions of charged particles from Au+Au collisions at the maximum RHIC energy, $\sqrt{s_{NN}}$ = 200 GeV. Phys. Rev. Lett. **88**, 202301 (2002)
18. B.B. Back et al., [PHOBOS Collaboration], Centrality dependence of the charged particle multiplicity near mid-rapidity in Au + Au collisions at $\sqrt{s_{NN}}$ = 130 GeV and 200 GeV. Phys. Rev. C **65**, 061901 (2002)
19. S.S. Adler et al., [PHENIX Collaboration], Systematic studies of the centrality and $\sqrt{s_{NN}}$ dependence of the $dE_T/d\eta$ and $dN_{ch}/d\eta$ in heavy ion collisions at mid-rapidity. Phys. Rev. C **71**, 034908 (2005)
20. S. Chatrchyan et al. [CMS Collaboration], Dependence on pseudorapidity and centrality of charged hadron production in PbPb collisions at a nucleon-nucleon centre-of-mass energy of 2.76 TeV. JHEP 1108, 141 (2011).
21. K. Aamodt et al., [ALICE Collaboration], Suppression of charged particle production at large transverse momentum in central Pb-Pb collisions at $\sqrt{s_{NN}}$ = 2.76 TeV. Phys. Lett. B **696**, 30 (2011)
22. A.M. Poskanzer, S.A. Voloshin, Methods for analyzing anisotropic flow in relativistic nuclear collisions. Phys. Rev. C **58**, 1671 (1998)
23. J. Takahashi, B.M. Tavares, W.L. Qian, R. Andrade, F. Grassi, Y. Hama, T. Kodama, N. Xu, Topology studies of hydrodynamics using two particle correlation analysis. Phys. Rev. Lett. **103**, 242301 (2009)
24. B. Alver, G. Roland, Collision geometry fluctuations and triangular flow in heavy-ion collisions. Phys. Rev. C 81, 054905 (2010) [Erratum-ibid. C 82, 039903 (2010)].
25. A. Adare et al., [PHENIX Collaboration], Measurements of higher-order flow harmonics in Au+Au collisions at $\sqrt{s_{NN}}$ = 200 GeV. Phys. Rev. Lett. **107**, 252301 (2011)
26. K. Aamodt et al., [ALICE Collaboration], Higher harmonic anisotropic flow measurements of charged particles in Pb-Pb collisions at $\sqrt{s_{NN}}$ = 2.76 TeV. Phys. Rev. Lett. **107**, 032301 (2011)
27. J. D. Bjorken, Energy loss of energetic partons in quark - gluon plasma: possible extinction of high p_T jets in hadron-hadron collisions, FERMILAB-PUB-82-059-THY (1982).
28. J. Adams et al., [STAR Collaboration], Evidence from d + Au measurements for final state suppression of high p_T hadrons in Au+Au collisions at RHIC. Phys. Rev. Lett. **91**, 072304 (2003)
29. C. Adler et al., [STAR Collaboration], Disappearance of back-to-back high p_T hadron correlations in central Au+Au collisions at $\sqrt{s_{NN}}$ = 200 GeV. Phys. Rev. Lett. **90**, 082302 (2003)
30. T. Matsui, H. Satz, J/psi Suppression by quark-gluon plasma formation. Phys. Lett. B **178**, 416 (1986)
31. S. Chatrchyan et al., [CMS Collaboration], Observation of sequential Upsilon suppression in PbPb collisions. Phys. Rev. Lett. **109**, 222301 (2012)

32. A. Adare et al., [PHENIX Collaboration], Enhanced production of direct photons in Au+Au collisions at $\sqrt{s_{NN}}$=200 GeV and implications for the initial temperature. Phys. Rev. Lett. **104**, 132301 (2010)
33. M. Wilde, Measurement of direct photons in pp and Pb-Pb collisions with ALICE. arXiv:1210.5958 [hep-ex] (2012)
34. A. Adare et al., [PHENIX Collaboration], Observation of direct-photon collective flow in $\sqrt{s_{NN}}$=200 GeV Au+Au collisions. Phys. Rev. Lett. **109**, 122302 (2012)
35. S. Chatrchyan et al., [CMS Collaboration], Study of Z boson production in PbPb collisions at nucleon-nucleon centre of mass energy = 2.76 TeV. Phys. Rev. Lett. **106**, 212301 (2011)
36. S. Chatrchyan et al., [CMS Collaboration], Study of W boson production in PbPb and pp collisions at sqrt(s[NN]) = 2.76 TeV. Phys. Lett. B **715**, 66 (2012)
37. L.V. Gribov, E.M. Levin, M.G. Ryskin, Semihard processes in QCD. Phys. Rept. **100**, 1 (1983)
38. A.H. Mueller, J.-W. Qiu, Gluonrecombination and shadowing at small values of x. Nucl. Phys. B **268**, 427 (1986)
39. J.P. Blaizot, A.H. Mueller, The early stage of ultrarelativistic heavy ion collisions. Nucl. Phys. B **289**, 847 (1987)
40. L.D. McLerran, R. Venugopalan, Computing quark and gluon distribution functions for very large nuclei. Phys. Rev. D **49**, 2233 (1994)
41. L.D. McLerran, R. Venugopalan, Gluon distribution functions for very large nuclei at small transverse momentum. Phys. Rev. D **49**, 3352 (1994)
42. L.D. McLerran, R. Venugopalan, Green's functions in the color field of a large nucleus. Phys. Rev. D **50**, 2225 (1994)
43. I. Balitsky, Operator expansion for high-energy scattering. Nucl. Phys. B **463**, 99 (1996)
44. Y.V. Kovchegov, Small x F(2) structure function of a nucleus including multiple pomeron exchanges. Phys. Rev. D **60**, 034008 (1999)
45. J. Jalilian-Marian, A. Kovner, L.D. McLerran, H. Weigert, The intrinsic glue distribution at very small x. Phys. Rev. D **55**, 5414 (1997)
46. J. Jalilian-Marian, A. Kovner, A. Leonidov, H. Weigert, The BFKL equation from the Wilson renormalization group. Nucl. Phys. B **504**, 415 (1997)
47. E. Iancu, A. Leonidov, L.D. McLerran, The renormalization group equation for the color glass condensate. Phys. Lett. B **510**, 133 (2001)
48. E. Iancu, A. Leonidov, L.D. McLerran, Nonlinear gluon evolution in the color glass condensate 1. Nucl. Phys. A **692**, 583 (2001)
49. E. Ferreiro, E. Iancu, A. Leonidov, L. McLerran, Nonlinear gluon evolution in the color glass condensate 2. Nucl. Phys. A **703**, 489 (2002)
50. F.D. Aaron et al., [H1 and ZEUS Collaboration], Combined measurement and QCD analysis of the inclusive $e^{\pm}p$ scattering cross sections at HERA. JHEP **1001**, 109 (2010)
51. K.J. Golec-Biernat, M. Wusthoff, Saturation effects in deep inelastic scattering at low Q^2 and its implications on diffraction. Phys. Rev. D **59**, 014017 (1998)
52. K.J. Golec-Biernat, M. Wusthoff, Saturation in diffractive deep inelastic scattering. Phys. Rev. D **60**, 114023 (1999)
53. D. Kharzeev, M. Nardi, Hadron production in nuclear collisions at RHIC and high density QCD. Phys. Lett. B **507**, 121 (2001)
54. D. Kharzeev, E. Levin, Manifestations of high density QCD in the first RHIC data. Phys. Lett. B **523**, 79 (2001)
55. D. Kharzeev, E. Levin, M. Nardi, The onset of classical QCD dynamics in relativistic heavy ion collisions. Phys. Rev. C **71**, 054903 (2005)
56. D. Kharzeev, E. Levin, M. Nardi, QCD saturation and deuteron nucleus collisions. Nucl. Phys. A 730, 448 (2004) [Erratum-ibid. A 743, 329 (2004)]
57. D. Kharzeev, E. Levin, M. Nardi, Color glass condensate at the LHC: hadron multiplicities in pp, pA and AA collisions. Nucl. Phys. A **747**, 609 (2005)
58. E. Levin, A.H. Rezaeian, Hadron multiplicity in pp and AA collisions at LHC from the color glass condensate. Phys. Rev. D **82**, 054003 (2010)

59. E. Levin, A.H. Rezaeian, Gluon saturation and energy dependence of hadron multiplicity in pp and AA collisions at the LHC. Phys. Rev. D **83**, 114001 (2011)
60. L. McLerran, M. Praszalowicz, Saturation and scaling of multiplicity, mean p_T and p_T distributions from 200 GeV $< \sqrt{s} <$ 7 TeV. Acta Phys. Polon. B **41**, 1917 (2010)
61. J.L. Albacete, Particle multiplicities in Lead-Lead collisions at the LHC from non-linear evolution with running coupling. Phys. Rev. Lett. **99**, 262301 (2007)
62. J. L. ALbacete, A. Dumitru, A model for gluon production in heavy-ion collisions at the LHC with rcBK unintegrated gluon densities. arXiv:1011.5161 [hep-ph] (2010)
63. J.L. Albacete, A. Dumitru, H. Fujii, Y. Nara, CGC predictions for p+Pb collisions at the LHC. Nucl. Phys. A **897**, 1 (2013)
64. A. Dumitru, D.E. Kharzeev, E.M. Levin, Y. Nara, Gluon saturation in pA collisions at the LHC: KLN model predictions for hadron multiplicities. Phys. Rev. C **85**, 044920 (2012)
65. C.-N. Yang, R.L. Mills, Conservation of isotopic spin and isotopic gauge invariance. Phys. Rev. **96**, 191 (1954)
66. T. Lappi, L. McLerran, Some features of the glasma. Nucl. Phys. A **772**, 200 (2006)
67. P. Romatschke, R. Venugopalan, Collective non-Abelian instabilities in a melting color glass condensate. Phys. Rev. Lett. **96**, 062302 (2006)
68. P. Romatschke, R. Venugopalan, The unstable glasma. Phys. Rev. D **74**, 045011 (2006)
69. H. Fujii, K. Itakura, Expanding color flux tubes and instabilities. Nucl. Phys. A **809**, 88 (2008)
70. A. Iwazaki, Decay of color gauge fields in heavy ion collisions and Nielsen-Olesen instability. Prog. Theor. Phys. **121**, 809 (2009)
71. H. Fujii, K. Itakura, A. Iwazaki, Instabilities in non-expanding glasma. Nucl. Phys. A **828**, 178 (2009)
72. K. Fukushima, F. Gelis, L. McLerran, Initial singularity of the little bang. Nucl. Phys. A **786**, 107 (2007)
73. J.-P. Blaizot, F. Gelis, J.-F. Liao, L. McLerran, R. Venugopalan, Bose-Einstein condensation and thermalization of the quark gluon plasma. Nucl. Phys. A **873**, 68 (2012)
74. L.D. Landau, S.Z. Belenkii, On the multiparticle production in high-energy collisions. Uspekhi Fiz. Nauk **56**, 309 (1955)
75. J.D. Bjorken, Highly relativistic nucleus-nucleus collisions: the central rapidity region. Phys. Rev. D **27**, 140 (1983)
76. P.F. Kolb, P. Huovinen, U.W. Heinz, H. Heiselberg, Elliptic flow at SPS and RHIC: from kinetic transport to hydrodynamics. Phys. Lett. B **500**, 232 (2001)
77. P. Huovinen, P.F. Kolb, U.W. Heinz, P.V. Ruuskanen, S.A. Voloshin, Radial and elliptic flow at RHIC: further predictions. Phys. Lett. B **503**, 58 (2001)
78. P.F. Kolb, U.W. Heinz, P. Huovinen, K.J. Eskola, K. Tuominen, Centrality dependence of multiplicity, transverse energy, and elliptic flow from hydrodynamics. Nucl. Phys. A **696**, 197 (2001)
79. D. Teaney, J. Lauret, E.V. Shuryak, Flow at the SPS and RHIC as a quark gluon plasma signature. Phys. Rev. Lett. **86**, 4783 (2001)
80. D. Teaney, J. Lauret, E. V. Shuryak, A hydrodynamic description of heavy ion collisions at the SPS and RHIC. arXiv:nucl-th/0110037 (2001)
81. T. Hirano, Is early thermalization achieved only near mid-rapidity at RHIC? Phys. Rev. C **65**, 011901 (2002)
82. T. Hirano, K. Tsuda, Collective flow and two pion correlations from a relativistic hydrodynamic model with early chemical freezeout. Phys. Rev. C **66**, 054905 (2002)
83. S.A. Bass, A. Dumitru, Dynamics of hot bulk QCD matter: from the quark gluon plasma to hadronic freezeout. Phys. Rev. C **61**, 064909 (2000)
84. T. Hirano, U.W. Heinz, D. Kharzeev, R. Lacey, Y. Nara, Hadronic dissipative effects on elliptic flow in ultrarelativistic heavy-ion collisions. Phys. Lett. B **636**, 299 (2006)
85. T. Hirano, U.W. Heinz, D. Kharzeev, R. Lacey, Y. Nara, Mass ordering of differential elliptic flow and its violation for phi mesons. Phys. Rev. C **77**, 044909 (2008)
86. C. Nonaka, S.A. Bass, Space-time evolution of bulk QCD matter. Phys. Rev. C **75**, 014902 (2007)

87. J. Steinheimer, M. Bleicher, H. Petersen, S. Schramm, H. Stocker, D. Zschiesche, (3+1)-dimensional hydrodynamic expansion with a critical point from realistic initial conditions. Phys. Rev. C **77**, 034901 (2008)
88. H. Petersen, J. Steinheimer, G. Burau, M. Bleicher, H. Stocker, A fully integrated transport approach to heavy ion reactions with an intermediate hydrodynamic stage. Phys. Rev. C **78**, 044901 (2008)
89. H. Petersen, G.-Y. Qin, S.A. Bass, B. Müller, Triangular flow in event-by-event ideal hydrodynamics in Au+Au collisions at $\sqrt{s_{\rm NN}} = 200A$ GeV. Phys. Rev. C **82**, 041901 (2010)
90. R. Baier, P. Romatschke, U.A. Wiedemann, Dissipative hydrodynamics and heavy ion collisions. Phys. Rev. C **73**, 064903 (2006)
91. R. Baier, P. Romatschke, U.A. Wiedemann, Transverse flow in relativistic viscous hydrodynamics. Nucl. Phys. A **782**, 313 (2007)
92. R. Baier, P. Romatschke, Causal viscous hydrodynamics for central heavy-ion collisions. Eur. Phys. J. C **51**, 677 (2007)
93. P. Romatschke, Causal viscous hydrodynamics for central heavy-ion collisions. II. Meson spectra and HBT radii. Eur. Phys. J. C **52**, 203 (2007)
94. P. Romatschke, U. Romatschke, Viscosity information from relativistic nuclear collisions: how perfect is the fluid observed at RHIC? Phys. Rev. Lett. **99**, 172301 (2007)
95. M. Luzum, P. Romatschke, Conformal relativistic viscous hydrodynamics: applications to RHIC results at $\sqrt{s_{NN}} = 200$ GeV. Phys. Rev. C 78, 034915 (2008) [Erratum-ibid. C 79, 039903 (2009)]
96. M. Luzum, P. Romatschke, Viscous hydrodynamic predictions for nuclear collisions at the LHC. Phys. Rev. Lett. **103**, 262302 (2009)
97. E. Retinskaya, M. Luzum, J.-Y. Ollitrault, Directed flow at midrapidity in $\sqrt{s_{NN}} = 2.76$ TeV Pb+Pb collisions. Phys. Rev. Lett. **108**, 252302 (2012)
98. A. K. Chaudhuri, Causal dissipative hydrodynamics for QGP fluid in 2+1 dimensions. arXiv:0704.0134 [nucl-th] (2007)
99. A. K. Chaudhuri, Saturation of elliptic flow and shear viscosity. arXiv:0708.1252 [nucl-th] (2007)
100. A. K. Chaudhuri, Viscous fluid dynamics in Au+Au collisions at RHIC. arXiv:0801.3180 [nucl-th] (2008)
101. A.K. Chaudhuri, Multiplicity, mean p_T, p_T-spectra and elliptic flow of identified particles in Pb+Pb collisions at LHC. Phys. Lett. B **672**, 126 (2009)
102. A.K. Chaudhuri, Initial eccentricity and constituent quark number scaling of elliptic flow in ideal and viscous dynamics. Phys. Rev. C **81**, 044905 (2010)
103. A.K. Chaudhuri, Nearly perfect fluid in Au+Au collisions at RHIC. Phys. Lett. B **681**, 418 (2009)
104. A.K. Chaudhuri, Knudsen number, ideal hydrodynamic limit for elliptic flow and QGP viscosity in $\sqrt{s}$=62 and 200 GeV Cu+Cu/Au+Au collisions. Phys. Rev. C **82**, 047901 (2010)
105. V. Roy, A.K. Chaudhuri, Hadronic resonance gas and charged particle's p_T spectra and elliptic flow in $\sqrt{s}$=200 GeV Au+Au collisions. Phys. Rev. C **82**, 031901 (2010)
106. A.K. Chaudhuri, V. Roy, Hydrodynamical analysis of centrality dependence of charged particle's multiplicity in $\sqrt{s}_{NN}$=2.76 TeV Pb+Pb collisions. Phys. Rev. C **84**, 027902 (2011)
107. V. Roy, A.K. Chaudhuri, Charged particle's elliptic flow in 2+1D viscous hydrodynamics at LHC ($\sqrt{s}$= 2.76 TeV) energy in Pb+Pb collision. Phys. Lett. B **703**, 313 (2011)
108. H. Song, U.W. Heinz, Suppression of elliptic flow in a minimally viscous quark-gluon plasma. Phys. Lett. B **658**, 279 (2008)
109. H. Song, U.W. Heinz, Causal viscous hydrodynamics in 2+1 dimensions for relativistic heavy-ion collisions. Phys. Rev. C **77**, 064901 (2008)
110. H. Song, U.W. Heinz, Multiplicity scaling in ideal and viscous hydrodynamics. Phys. Rev. C **78**, 024902 (2008)
111. U.W. Heinz, J.S. Moreland, H. Song, Viscosity from elliptic flow: the Path to precision. Phys. Rev. C **80**, 061901 (2009)

112. K. Dusling, D. Teaney, Simulating elliptic flow with viscous hydrodynamics. Phys. Rev. C **77**, 034905 (2008)
113. G. Torrieri, I. Mishustin, Instability of boost-invariant hydrodynamics with a QCD inspired bulk viscosity. Phys. Rev. C **78**, 021901 (2008)
114. R.J. Fries, B. Müller, A. Schäfer, Stress tensor and bulk viscosity in relativistic nuclear collisions. Phys. Rev. C **78**, 034913 (2008)
115. G.S. Denicol, T. Kodama, T. Koide, P. Mota, Effect of bulk viscosity on elliptic flow near QCD phase transition. Phys. Rev. C **80**, 064901 (2009)
116. H. Song, U.W. Heinz, Interplay of shear and bulk viscosity in generating flow in heavy-ion collisions. Phys. Rev. C **81**, 024905 (2010)
117. P. Bozek, Bulk and shear viscosities of matter created in relativistic heavy-ion collisions. Phys. Rev. C **81**, 034909 (2010)
118. V. Roy, A. K. Chaudhuri, 2+1 dimensional hydrodynamics including bulk viscosity: a systematics study. Phys. Rev. C 85, 024909 (2012) [Erratum-ibid. C 85, 049902 (2012)]
119. C. Shen, U. Heinz, P. Huovinen, H. Song, Systematic parameter study of hadron spectra and elliptic flow from viscous hydrodynamic simulations of Au+Au collisions at $\sqrt{s_{NN}} = 200$ GeV. Phys. Rev. C **82**, 054904 (2010)
120. C. Shen, U. Heinz, Collision energy dependence of viscous hydrodynamic flow in relativistic heavy-ion collisions. Phys. Rev. C 85, 054902 (2012) [Erratum-ibid. C 86, 049903 (2012)]
121. B.H. Alver, C. Gombeaud, M. Luzum, J.-Y. Ollitrault, Triangular flow in hydrodynamics and transport theory. Phys. Rev. C **82**, 034913 (2010)
122. Z. Qiu, C. Shen, U. Heinz, Hydrodynamic elliptic and triangular flow in Pb-Pb collisions at $\sqrt{s} = 2.76$ATeV. Phys. Lett. B **707**, 151 (2012)
123. A.K. Chaudhuri, Influence of shear viscosity on the correlation between the triangular flow and initial spatial triangularity. Phys. Lett. B **713**, 91 (2012)
124. M.Rihan Haque, V. Roy, A. K. Chaudhuri, Fluctuating initial condition and smoothening effect on elliptic and triangular flow. Phys. Rev. C 86, 037901 (2012)
125. A. K. Chaudhuri, M.R. Haque, V. Roy, B. Mohanty, Even-by-event hydrodynamical simulations for $\sqrt{s}_{NN}$=200 GeV Au+Au collisions and the correlation between flow coefficients and initial asymmetry measures. arXiv:1211.2040 [nucl-th] (2012)
126. H. Song, S. A. Bass, U. Heinz, T. Hirano, C. Shen, 200 A GeV Au+Au collisions serve a nearly perfect quark-gluon liquid. Phys. Rev. Lett. 106, 192301 (2011) [Erratum-ibid. 109, 139904 (2012)]
127. H. Song, S. A. Bass, U. Heinz, T. Hirano, C. Shen, Hadron spectra and elliptic flow for 200 A GeV Au+Au collisions from viscous hydrodynamics coupled to a Boltzmann cascade. Phys. Rev. C 83, 054910 (2011) [Erratum-ibid. C 86, 059903 (2012)]
128. H. Song, S.A. Bass, U. Heinz, Viscous QCD matter in a hybrid hydrodynamic+Boltzmann approach. Phys. Rev. C **83**, 024912 (2011)
129. P. Bozek, Viscous evolution of the rapidity distribution of matter created in relativistic heavy-ion collisions. Phys. Rev. C **77**, 034911 (2008)
130. A. Monnai, T. Hirano, Longitudinal viscous hydrodynamic evolution for the shattered colour glass condensate. Phys. Lett. B **703**, 583 (2011)
131. A. Monnai, Dissipative hydrodynamic effects on baryon stopping. Phys. Rev. C **86**, 014908 (2012)
132. B. Schenke, S. Jeon, C. Gale, Elliptic and triangular flow in event-by-event (3+1)D viscous hydrodynamics. Phys. Rev. Lett. **106**, 042301 (2011)
133. B. Schenke, S. Jeon, C. Gale, Anisotropic flow in sqrt(s)=2.76 TeV Pb+Pb collisions at the LHC. Phys. Lett. B **702**, 59 (2011)
134. B. Schenke, S. Jeon, C. Gale, Higher flow harmonics from (3+1)D event-by-event viscous hydrodynamics. Phys. Rev. C **85**, 024901 (2012)
135. P. Bozek, Flow and interferometry in 3+1 dimensional viscous hydrodynamics. Phys. Rev. C **85**, 034901 (2012)
136. P. Bozek, I. Wyskiel-Piekarska, Particle spectra in Pb-Pb collisions at $sgrts_{NN}$ = 2.76 TeV. Phys. Rev. C **85**, 064915 (2012)

137. P. Bozek, Event-by-event viscous hydrodynamics for Cu-Au collisions at 200GeV. Phys. Lett. B **717**, 287 (2012)
138. J. Vredevoogd, S. Pratt, Viscous hydrodynamics and relativistic heavy ion collisions. arXiv:1202.1509 [nucl-th] (2012)
139. H. Niemi, G.S. Denicol, P. Huovinen, E. Molnar, D.H. Rischke, Influence of the shear viscosity of the quark-gluon plasma on elliptic flow in ultrarelativistic heavy-ion collisions. Phys. Rev. Lett. **106**, 212302 (2011)
140. H. Niemi, G.S. Denicol, P. Huovinen, E. Molnar, D.H. Rischke, Influence of a temperature-dependent shear viscosity on the azimuthal asymmetries of transverse momentum spectra in ultrarelativistic heavy-ion collisions. Phys. Rev. C **86**, 014909 (2012)
141. C. Shen, U. Heinz, P. Huovinen, H. Song, Radial and elliptic flow in Pb+Pb collisions at the Large Hadron Collider from viscous hydrodynamic. Phys. Rev. C **84**, 044903 (2011)
142. M. Martinez, M. Strickland, Dissipative dynamics of highly anisotropic systems. Nucl. Phys. A **848**, 183 (2010)
143. M. Martinez, M. Strickland, Non-boost-invariant anisotropic dynamics. Nucl. Phys. A **856**, 68 (2011)
144. W. Florkowski, R. Ryblewski, Highly-anisotropic and strongly-dissipative hydrodynamics for early stages of relativistic heavy-ion collisions. Phys. Rev. C **83**, 034907 (2011)
145. R. Ryblewski, W. Florkowski, Highly-anisotropic and strongly-dissipative hydrodynamics with transverse expansion. Eur. Phys. J. C **71**, 1761 (2011)
146. J.I. Kapusta, B. Müller, M. Stephanov, Relativistic theory of hydrodynamic fluctuations with applications to heavy ion collisions. Phys. Rev. C **85**, 054906 (2012)
147. D. Teaney, L. Yan, Triangularity and dipole asymmetry in heavy ion collisions. Phys. Rev. C **83**, 064904 (2011)
148. D. Teaney, L. Yan, Non linearities in the harmonic spectrum of heavy ion collisions with ideal and viscous hydrodynamics. Phys. Rev. C **86**, 044908 (2012)
149. Z. Qiu, U. Heinz, Hydrodynamic event-plane correlations in Pb+Pb collisions at $\sqrt{s}$ = 2.76ATeV. Phys. Lett. B **717**, 261 (2012)
150. H.-J. Drescher, Y. Nara, Effects of fluctuations on the initial eccentricity from the color glass condensate in heavy ion collisions. Phys. Rev. C **75**, 034905 (2007)
151. H.-J. Drescher, Y. Nara, Eccentricity fluctuations from the color glass condensate at RHIC and LHC. Phys. Rev. C **76**, 041903 (2007)
152. B. Schenke, P. Tribedy, R. Venugopalan, Fluctuating glasma initial conditions and flow in heavy ion collisions. Phys. Rev. Lett. **108**, 252301 (2012)
153. C. Gale, S. Jeon, B. Schenke, P. Tribedy, R. Venugopalan, Event-by-event anisotropic flow in heavy-ion collisions from combined Yang-Mills and viscous fluid dynamics. arXiv:1209.6330 [nucl-th] (2012)
154. F. Cooper, G. Frye, Comment On the single particle distribution in the hydrodynamic and statistical thermodynamic models Of multiparticle production. Phys. Rev. D **10**, 186 (1974)
155. D. Teaney, Effect of shear viscosity on spectra, elliptic flow, and Hanbury Brown-Twiss radii. Phys. Rev. C **68**, 034913 (2003)
156. A. Monnai, T. Hirano, Effects of bulk viscosity at freezeout. Phys. Rev. C **80**, 054906 (2009)
157. K. Dusling, T. Schafer, Phys. Rev. C **85**, 044909 (2012)
158. Y. Nara, N. Otuka, A. Ohnishi, K. Niita, S. Chiba, Study of relativistic nuclear collisions at AGS energies from p + Be to Au + Au with hadronic cascade model. Phys. Rev. C **61**, 024901 (2000)
159. S.A. Bass, M. Belkacem, M. Bleicher, M. Brandstetter, L. Bravina, C. Ernst, L. Gerland, M. Hofmann et al., Microscopic models for ultrarelativistic heavy ion collisions. Prog. Part. Nucl. Phys. **41**, 255 (1998)
160. M. Bleicher, E. Zabrodin, C. Spieles, S.A. Bass, C. Ernst, S. Soff, L. Bravina, M. Belkacem et al., Relativistic hadron hadron collisions in the ultrarelativistic quantum molecular dynamics model. J. Phys. G 25, 1859 (1999)

Chapter 3
Relativistic Dissipative Hydrodynamics with Conserved Charges

Hydrodynamics is a generic theoretical framework to describe the collective motions of strongly-interacting many-body systems [1]. Microscopic details of the medium of interest are coarse grained in hydrodynamics—for example, the average motion of the constituent particles is denoted by the flow. Generally speaking, an arbitrary system needs many-body distribution function to describe them. When there are interactions among the constituent particles, it would enter the stage where the one-body distribution function is sufficient for understanding the system. It is called the kinetic stage, and this happens typically roughly at the time scale of the collision time. As the system equilibrates in time evolution, it would eventually be locally thermalized and reach the hydrodynamic stage. Here only a limited number of thermodynamic variables are necessary for the description of the medium. The thermodynamic variables are defined at each point on the space-time. This implies that one considers two different scales in terms of volume for hydrodynamics; an infinite volume is required for thermodynamic variables to be well-defined whereas the size of this thermodynamic cell must be small compared with the size of the whole system.

Its relativistic version, *relativistic hydrodynamics* [2], is considered to describe various phenomena from interstellar gas to the quark-gluon plasma. It is known that relativistic ideal hydrodynamics can be formulated as a straight-forward extension of the non-relativistic one. On the other hand, the formulation of relativistic dissipative hydrodynamics, which takes account of viscous and dissipative processes, involve many non-trivialities, because off-equilibrium corrections cannot be described by the relativistic Navier–Stokes equation as it suffers from acausality and instability to infinitesimal perturbation from equilibrium [3]. Thus one conventionally introduces relaxation effects to fix those problems [4]. Various formalisms of causal hydrodynamics have been proposed so far [5–27].

The theories of relativistic dissipative hydrodynamics so far explicitly assumes binary collisions with particle number conservation, including Israel-Stewart theory [5, 6], one of the most widely accepted causal hydrodynamic theory in viscous hydrodynamic analyses of the hot QCD matter. On the other hand, particle number changing processes would be essential in high-energy heavy ion collisions both for

A. Monnai, *Relativistic Dissipative Hydrodynamic Description of the Quark-Gluon Plasma*, Springer Theses, DOI: 10.1007/978-4-431-54798-3_3,

the hadronic and the QGP phases. In this chapter, a systematic formulation of causal dissipative hydrodynamics for the systems with multiple conserved charges—and consequently with multiple components—is presented [28]. It is found that linear cross terms, which has been missing in the previous relativistic formalisms, play an important role in the consistent formulation.

3.1 Relativistic Ideal Hydrodynamics

Ideal hydrodynamics is defined as a special case of dissipative hydrodynamics, where the transport coefficients are identically zero. It is important to keep in mind this does not necessarily mean *global* equilibrium where the thermodynamic forces do not exist in the system at all, because entropy is not produced if the energy-momentum tensor and the conserved charge currents just do not respond to the forces.

Relativistic hydrodynamics introduces the flow $u^\mu(x)$ which is defined as a 4-velocity, i.e., $u^\mu = \gamma(1, v_x, v_y, v_z)$ where γ is the Lorentz dilation factor;

$$\gamma = \frac{1}{\sqrt{1 - v_x^2 - v_y^2 - v_z^2}}. \tag{3.1}$$

The flow is normalized as $u^\mu u_\mu = 1$. The system is essentially described by the energy-momentum tensor $T^{\mu\nu}(x)$ and the conserved charge current $N_J^\mu(x)$ ($J = 1, \ldots, N$). The temperature $T(x)$ and the chemical potentials $\mu_J(x)$ can be derived from those conserved quantities using the thermodynamic relations and the equation of state. The energy-momentum tensor and the Jth conserved currents in an ideal hydrodynamic system are expressed as

$$T_0^{\mu\nu} = e_0 u^\mu u^\nu - P_0 \Delta^{\mu\nu}, \tag{3.2}$$

$$N_{J0}^\mu = n_{J0} u^\mu, \tag{3.3}$$

where the spatial projection operator $\Delta^{\mu\nu} = g^{\mu\nu} - u^\mu u^\nu$ is defined with the Minkowski metric $g^{\mu\nu} = \mathrm{diag}(+, -, -, -)$. It is note-worthy that the orthogonality relation $\Delta^{\mu\nu} u_\nu = 0$ is satisfied. $e_0 = u_\mu u_\nu T_0^{\mu\nu}$, $P_0 = -\frac{1}{3}\Delta_{\mu\nu} T_0^{\mu\nu}$ and $n_{J0} = u_\mu N_{J0}^\mu$ are the energy density, the hydrostatic pressure and the conserved charge density in the local rest frame. One recovers the traditional single component system with particle number conservation by choosing $N = 1$ and considering the particle number to be the conserved charge.

They satisfy the conservation laws[1] which read

$$\partial_\mu T^{\mu\nu} = \partial_\mu T_0^{\mu\nu} = 0, \tag{3.4}$$

$$\partial_\mu N_J^\mu = \partial_\mu N_{J0}^\mu = 0. \tag{3.5}$$

[1] One may consider quantum anomaly in the presence of external fields. See, for example, Ref. [29] for the discussion.

These equations provide $4 + N$ independent equations for describing the system. Since the number of independent variables is $5 + N$ for e_0, P_0, n_{J0} and u^μ, the conservation equations alone do not give complete description. This indicate that microscopic physics has to provide the remaining one relation for these quantities to characterize the system. One usually employs the equation of state $P_0 = P_0(e_0, \{n_{J0}\})$ for this purpose.

The entropy current is given as $s^\mu = su^\mu$ where the entropy density is $s = (e_0 + P_0 - \mu_J n_{J0})/T$. One can see the entropy is not produced in the system as

$$\begin{aligned}\partial_\mu s^\mu &= \frac{1}{T} u_\nu \partial_\mu T_0^{\mu\nu} + \frac{\mu_J}{T} \partial_\mu N_{J0}^\mu + u^\mu \left[\frac{1}{T} \partial_\mu P_0 + (e_0 + P_0) \partial_\mu \frac{1}{T} - n_{J0} \partial_\mu \frac{\mu_J}{T} \right] \\ &= 0, \end{aligned} \tag{3.6}$$

identically holds due to the conservation laws and the Gibbs–Duhem relation (see Appendix B for details). It should be noted that energy and the number densities still transmit through the medium with the flow as convection.

3.2 Relativistic Dissipative Hydrodynamics

It is implied from the uncertainty principle that viscosity does not truly vanish, i.e., there would not be a perfect liquid. The finiteness of shear viscosity is discussed in the contexts of kinetic theory [30] and the AdS/CFT correspondence [31]. Thus it is of particular importance to take into account dissipative processes for the realistic description of the physical phenomena of interest. Dissipative hydrodynamics describes non-equilibrium processes of the target system as corrections to ideal hydrodynamics. The energy-momentum tensor and the conserved charge current can be expressed as

$$T^{\mu\nu} = (e_0 + \delta e) u^\mu u^\nu - (P_0 + \Pi) \Delta^{\mu\nu} + W^\mu u^\nu + W^\nu u^\mu + \pi^{\mu\nu}, \tag{3.7}$$

$$N_J^\mu = (n_{J0} + \delta n_J) u^\mu + V_J^\mu, \tag{3.8}$$

where $\Pi = -\frac{1}{3}\Delta_{\mu\nu}\delta T^{\mu\nu}$ is the bulk pressure, $W^\mu = u_\nu \Delta^{\mu\rho} \delta T_\rho^\nu$ is the energy dissipation current, $\pi^{\mu\nu} = [\frac{1}{2}(\Delta^{\mu\alpha}\Delta^{\nu\beta} + \Delta^{\nu\alpha}\Delta^{\mu\beta}) - \frac{1}{3}\Delta^{\mu\nu}\Delta_{\alpha\beta}]\delta T^{\mu\nu} = \delta T^{\langle\mu\nu\rangle}$ is the shear stress tensor and $V_J^\mu = \Delta^{\mu\nu} N_\nu^J$ is the Jth charge dissipation current. $\delta e = u_\mu u_\nu \delta T^{\mu\nu}$ and $\delta n_J = u_\mu \delta N_J^\mu$ are the off-equilibrium corrections to the energy density and the charge density. It is found later that they should vanish for the thermodynamic stability of a system. This is known as Landau matching condition, though it is originally meant as a phenomenological constraint to reduce the number of variables rather that a thermodynamic requirement. The entropy production no longer vanishes because the non-equilibrium components of the conservation laws now appears in Eq. (3.6) and so does that of the entropy current itself.

In non-relativistic fluids, heat is transmitted via conduction, convection and radiation. Thermal conduction is a pure transmission of heat which does not involve the movement of the constituent particles. This is related with the heat dissipation defined as

$$q^\mu = W^\mu - \sum_J \frac{e_0 + P_0}{n_{J0}} V_J^\mu, \tag{3.9}$$

in relativistic dissipative hydrodynamics. Thermal convection, on the other hand, is induced via the transportation of the constituent particles themselves. This is embodied by the flow u^μ and associated energy density transfer. Radiation corresponds to thermal photon emission and absorption, even though the latter might be small since the medium is electromagnetically transparent. The jet of colored objects might be interpreted as radiation that can contribute to the medium.

The additional set of macroscopic variables introduced in dissipative hydrodynamics increases the number of variables by $10 + 4N$. They cannot be determined from the conservation laws and the equation of state alone since the dissipative processes represent higher order response of the medium, which is dependent on microscopic properties of the medium. Thus one has to have the equations of motions for the dissipative currents—sometimes referred to as constitutive equations from phenomenological perspective—by introducing the corresponding transport coefficients.

From now on the term *viscous* hydrodynamics is used to refer to the hydrodynamics with shear and bulk viscosities, which are the tensor and the scalar off-equilibrium processes, and *dissipative* hydrodynamics to the one with charge dissipations, the vector processes, in addition to the viscosities.

3.2.1 Local Rest Frame

The local rest frame is where the thermodynamic variables are defined. The flow has the time-like component only, i.e., $u^\mu_{\mathrm{LRF}} = (1, 0, 0, 0)$ in this frame. Likewise the projection operator only has the space-like components: $\Delta^{\mu\nu}_{\mathrm{LRF}} = \mathrm{diag}(0, -, -, -)$. It is related to the laboratory frame via the Lorentz transformation $\Lambda^\mu{}_\nu(x)$ as $u^\mu_{\mathrm{LRF}} = \Lambda^\mu{}_\nu u^\nu$. Then the energy-momentum tensor and the charge currents are expressed as

$$\begin{aligned} T^{\mu\nu} &= T_0^{\mu\nu} + \delta T^{\mu\nu} \\ &= \begin{pmatrix} e_0 & 0 & 0 & 0 \\ 0 & P_0 & 0 & 0 \\ 0 & 0 & P_0 & 0 \\ 0 & 0 & 0 & P_0 \end{pmatrix} + \begin{pmatrix} \delta e & W^x & W^y & W^z \\ W^x & \Pi + \pi^{xx} & \pi^{xy} & \pi^{xz} \\ W^y & \pi^{yx} & \Pi + \pi^{yy} & \pi^{yz} \\ W^z & \pi^{zx} & \pi^{zy} & \Pi + \pi^{zz} \end{pmatrix}, \\ N^\mu &= N_0^\mu + \delta N^\mu \end{aligned} \tag{3.10}$$

$$= \begin{pmatrix} n_{J0} \\ 0 \\ 0 \\ 0 \end{pmatrix} + \begin{pmatrix} \delta n_J \\ V_J^x \\ V_J^y \\ V_J^z \end{pmatrix}. \tag{3.11}$$

In the ideal hydrodynamic limit, the local rest frame satisfies $T^{\mu\nu}u_\nu = e_0 u^\mu$ and $N_J^\mu = n_{J_0}u^\mu$ so that there are no leaks from the local flux, i.e., the flow can be uniquely defined. In dissipative hydrodynamics, it is not a priori clear how to set the flow as the two relations cannot be satisfied simultaneously in generic systems. In principle one can choose an arbitrary frame that is not too far from the equilibrium hydrodynamic flow, since the only condition is that the dissipative components of the conserved quantities be small. There are, however, two common choices of the frame for defining the flow. Landau frame, or the energy frame, is the local rest frame of the energy flow defined as $T^{\mu\nu}u_\nu = e_0 u^\mu$. In this frame, the formalism has vanishing energy dissipation $W^\mu = 0$ and finite charge dissipation $V^\mu \neq 0$. Alternatively one may choose Eckart frame, or the particle/charge frame, which is the local rest frame of the conserved current. The flow is defined as $N_J^\mu = n_{J_0}u^\mu$ in this frame. For several reasons the former frame is preferred in relativistic hydrodynamics; firstly, the particle frame cannot be defined for the systems without conserved currents while the energy frame is always definable. Secondly, if there are multiple conserved currents, choosing one of the particle frames does not have the benefit of simplifying the formalism since the other currents do not vanish. It is argued in theoretical analyses (see for example, Ref. [32]) that the energy frame is better defined from the view point of hydrodynamic stability.

3.3 Derivation of Causal Dissipative Hydrodynamic Equations

Relativistic dissipative hydrodynamic equations are derived for multi-component systems with multiple conserved currents by extending the Israel-Stewart second order theory [6]. The formulation is based on relativistic kinetic theory (see Appendix C). In their method, the third moment of the phase-space distribution function is introduced and dissipative hydrodynamic equations are derived from its space-time derivative, which can be constrained by the second law of thermodynamics. In this chapter it will be shown that new moment equations are required to consistently describe the systems with multiple conserved currents.

Several non-trivialities of multi-component systems are found in the course of formulation. Firstly, the thermodynamic stability conditions (Appendix D), which ensure that the system is in maximum entropy state in terms of dissipative currents, have to be considered after the hydrodynamic equations of motion for the dissipative currents are obtained, because the number of the equations and that of the dissipative currents would not match if the conditions were used beforehand. Secondly, the second law of thermodynamics requires a specific form of moment expansion for the distortion of the distribution in multi-component systems, which also justifies the result of Ref. [33] developed in the context of freeze-out (see Appendix A for details). Thirdly, if the system has conserved charges, new moment equations are needed for consistent formulation of relativistic dissipative hydrodynamics. These equations also allow one to determine all the dissipative currents in an arbitrary frame, which the conventional Israel-Stewart theory cannot do.

The existence of multiple conserved currents also brings uncertainties to the conventional Grad's 14-moment method because it is no longer applicable when more than 14 dissipative currents are present. Systems with conservations based on quantum numbers, such as baryon number and electric charge is considered here, i.e., inelastic scatterings and chemical interactions are present. Here a generalized moment method which is quite in agreement with Onsager reciprocal relations [34, 35] is proposed to describe such systems without ambiguity.

3.3.1 First-Order Theory

The equations of motion for the dissipative currents are constrained by the law of increasing entropy, the macroscopic representation of irreversible processes. When the off-equilibrium deviation is small, it can be treated as perturbation in terms of the dissipative currents. By taking into account the correction to the entropy current up to the first order, one obtains *first-order theory*, which is also known as linear response theory. This corresponds to the relativistic version of Navier–Stokes equation. Here first order dissipative hydrodynamics for relativistic multi-component systems with multiple conserved currents is discussed from relativistic kinetic theory with emphasis on the Onsager reciprocal relation.

The energy momentum tensor $T^{\mu\nu}$, the conserved currents N_J^μ and the entropy current s^μ are expressed in kinetic theory with the phase-space distribution f^i as

$$T^{\mu\nu} = \sum_i \int \frac{g_i d^3 p}{(2\pi)^3 E_i} p_i^\mu p_i^\nu f^i, \tag{3.12}$$

$$N_J^\mu = \sum_i \int \frac{q_i^J g_i d^3 p}{(2\pi)^3 E_i} p_i^\mu f^i, \tag{3.13}$$

$$s^\mu = -\sum_i \int \frac{g_i d^3 p}{(2\pi)^3 E_i} p_i^\mu \phi(f^i), \tag{3.14}$$

where

$$\phi(f^i) = f^i \ln f^i - \varepsilon^{-1}(1+\varepsilon f^i)\ln(1+\varepsilon f^i). \tag{3.15}$$

Here g_i is the degeneracy and q_i^J is the quantum number for the Jth conserved current. The index i denotes the particle species ($i = 1, \dots, n$). The sign factor ε is $+1$ for bosons, -1 for fermions and 0 for classical particles. In the classical limit, $\phi(f^i)$ reduces to $f^i \ln f^i - f^i$. y_0^i is defined in the equilibrium distribution $f_0^i = [\exp(y_0^i) - \varepsilon]^{-1}$, which means

$$y_0^i = -\sum_J \frac{q_i^J \mu_J}{T} + p_i^\mu \frac{u_\mu}{T}. \tag{3.16}$$

The entropy current is expanded in terms of the distortion of distribution $\delta f^i = f^i - f_0^i$ as

$$\begin{aligned} s^\mu &= s_0^\mu + \delta s_{(1)}^\mu + \delta s_{(2)}^\mu + \dots \\ &= s_0^\mu + \sum_i \int \frac{g_i d^3 p}{(2\pi)^3 E_i} p_i^\mu y_0^i \delta f^i + \mathscr{O}(\delta f^2), \end{aligned} \tag{3.17}$$

up to the first order. Note that the number of conserved currents and that of particle species are generally different when inelastic processes are present. The entropy production is written as

$$\begin{aligned} \partial_\mu s^\mu &= \sum_i \int \frac{g_i d^3 p}{(2\pi)^3 E_i} p_i^\mu \delta f^i \partial_\mu y_0^i + \mathscr{O}[\partial(\delta f^2)] \\ &\simeq \delta T^{\mu\nu} \partial_\mu \frac{u_\nu}{T} - \sum_J \delta N_J^\mu \partial_\mu \frac{\mu_J}{T} \\ &= \delta e D \frac{1}{T} - \Pi \frac{1}{T} \nabla_\mu u^\mu + W^\mu \left(\nabla_\mu \frac{1}{T} + \frac{1}{T} D u_\mu \right) \\ &\quad + \pi^{\mu\nu} \frac{1}{T} \nabla_{\langle\mu} u_{\nu\rangle} - \sum_J \delta n_J D \frac{\mu_J}{T} - \sum_J V_J^\mu \nabla_\mu \frac{\mu_J}{T}, \end{aligned} \tag{3.18}$$

where the time-like and the space-like derivatives are defined as $D = u^\mu \partial_\mu$ and $\nabla_\mu = \Delta_{\mu\nu} \partial^\nu$, respectively. Energy-momentum conservation and charge number conservations are used in the equation above. Then one finds that the dissipative currents in linear response theory should be expressed as

$$\delta e = \zeta_{\delta e \delta e} D \frac{1}{T} - \sum_J \zeta_{\delta e \delta n_J} D \frac{\mu_J}{T} + \zeta_{\delta e \Pi} \frac{1}{T} \nabla_\mu u^\mu, \tag{3.19}$$

$$\Pi = -\zeta_{\Pi\Pi} \frac{1}{T} \nabla_\mu u^\mu - \zeta_{\Pi \delta e} D \frac{1}{T} + \sum_J \zeta_{\Pi \delta n_J} D \frac{\mu_J}{T}, \tag{3.20}$$

$$W^\mu = -\kappa_{WW} \left(\nabla^\mu \frac{1}{T} + \frac{1}{T} D u^\mu \right) + \sum_J \kappa_{W V_J} \nabla^\mu \frac{\mu_J}{T}, \tag{3.21}$$

$$\pi^{\mu\nu} = 2\eta_{\pi\pi}\frac{1}{T}\nabla^{\langle\mu}u^{\nu\rangle}, \tag{3.22}$$

$$\delta n_J = -\sum_K \zeta_{\delta n_J \delta n_K} D\frac{\mu_K}{T} + \zeta_{\delta n_J \Pi}\frac{1}{T}\nabla_\mu u^\mu + \zeta_{\delta n_J \delta e} D\frac{1}{T}, \tag{3.23}$$

$$V_J^\mu = \sum_K \kappa_{V_J V_K}\nabla^\mu \frac{\mu_K}{T} - \kappa_{V_J W}\left(\nabla^\mu \frac{1}{T} + \frac{1}{T}Du^\mu\right), \tag{3.24}$$

so the semi-positive definiteness of the entropy production is always satisfied. $\zeta = \zeta_{\Pi\Pi}/T$ is the bulk viscosity, $\kappa_W = \kappa_{WW}$ is the energy conductivity, $\eta = \eta_{\pi\pi}/T$ is the shear viscosity and $\kappa_{V_J} = \kappa_{V_J V_J}$ is the charge conductivity of the Jth conserved current. Note that different thermodynamic forces of the same tensor order are allowed in the constitutive equations, such as the gradient of the chemical potential over temperature μ_J/T in the constitutive equation for W^μ, even though they are often neglected in relativistic formulation. The cross transport coefficients should satisfy Onsager reciprocal relations [34], e.g., $\kappa_{WV_J} = \kappa_{V_J W}$ and have to be so chosen that semi-positive definiteness of the entropy production is preserved.

It might seem that each term in Eq. (3.18) should be expressed as quadratic forms in dissipative currents to obey the second law of thermodynamics, e.g., $W^\mu \propto (\nabla_\mu \frac{1}{T} + \frac{1}{T}Du_\mu)$ and some of the cross terms seemingly violate it when Onsager reciprocal relation is considered. However one can prove that the coefficients can be chosen so that the law of increasing entropy is satisfied. If the dissipative currents of the same tensor order are denoted as J_p and the corresponding thermodynamic forces as X_p where p denotes the type of the dissipative current, then the entropy production for the tensor order is

$$\partial_\mu s^\mu = \sum_p J_p X_p. \tag{3.25}$$

The dissipative currents in linear response theory are expressed as $J_p = \sum_q C_{pq}X_q$ where $C_{pq} = C_{qp}$ is an element of the transport coefficient matrix. Thus the entropy production can be expressed in a bilinear form as

$$\partial_\mu s^\mu = \sum_{p,q} X_p C_{pq} X_q. \tag{3.26}$$

On the other hand, one can diagonalize any symmetric matrix with a certain orthogonal matrix $P^{-1} = P^{\mathbf{T}}$ as $C_{pq} = \sum_{r,s} P_{pr}(C'_r\delta_{rs})P_{sq}$. Here one requires the diagonal elements C'_r to be semi-positive. Then

$$\begin{aligned}\partial_\mu s^\mu &= \sum_{p,q,r,s} X_p P_{pr}(C'_r\delta_{rs})P_{sq}X_q \\ &= \sum_{r,s} X'_r(C'_r\delta_{rs})X'_s\end{aligned}$$

$$= \sum_r C'_r X'^2_r \geq 0, \tag{3.27}$$

is obtained where $X'_p = \sum_q P_{pq} X_q$. Therefore if the appropriate linear combinations of the thermodynamic forces are chosen, the entropy production can be expressed in quadratic forms, which naturally satisfies the semi-positive definite condition.

It should be noted that the constitutive equations are in the form of linear response to the thermodynamic force, i.e., the derivatives of flow, temperature and chemical potentials and are independent of the specific form of the off-equilibrium distribution f^i. The constitutive equations for δe and δn_J are not written because they vanish when the thermodynamic stability conditions are employed. The details can be found in Appendix D.

The first-order theory, however, are known to be acausal and unstable [3] because it allows instantaneous propagation and also has exponentially growing modes against infinitesimal perturbation from global equilibrium. Those facts encourage one to consider the second order corrections to s^μ to introduce relaxation effects and obtain a causal theory.

3.3.2 Second-Order Theory

The second-order theory is an effective theory which evades the problems for causality and stability against infinitesimal perturbation. It is called "second-order", because it corresponds to introducing quadratic corrections to the dissipative current expansion of the entropy production. In this way the exact expression of the non-equilibrium variables require the time and space derivatives of the fields δN^μ_J and $\delta T^{\mu\nu}$, in addition to the linear order terms which involves the derivatives of μ_J, T and u^μ. The time derivative of the dissipative current itself in the equation implies that this can be interpreted as the equation of motion. The time evolution of the system is characterized by the relaxation time τ_J. The second order corrections are generally small, but since initial conditions are often given as boundary conditions, the relaxation effect plays an important role in the early stages of time evolution. Appendix F shows that the linear response form is recovered after the typical timescale of $\tau \sim \tau_J$.

Energy-momentum conservation (3.7) and charge number conservations (3.8) in non-equilibrium systems are tensor-decomposed into time-like and space-like components:

$$\begin{aligned} D(e_0 + \delta e) &= -(e_0 + \delta e + P_0 + \Pi)\nabla_\mu u^\mu \\ &\quad + 2W^\mu Du_\mu - \nabla_\mu W^\mu + \pi^{\mu\nu}\nabla_{\langle\mu}u_{\nu\rangle}, & (3.28) \\ (e_0 + \delta e + P_0 + \Pi)Du^\mu &= \nabla^\mu(P_0 + \Pi) - W^\mu\nabla_\nu u^\nu - \Delta^{\mu\nu}DW_\nu \\ &\quad - W^\nu\nabla_\nu u^\mu + \pi^{\mu\nu}Du_\nu - \Delta^{\mu\nu}\nabla^\rho\pi_{\nu\rho}, & (3.29) \\ D(n_{J0} + \delta n_J) &= -(n_{J0} + \delta n_J)\nabla_\mu u^\mu - \nabla_\mu V^\mu_J + V^\mu_J Du_\mu. & (3.30) \end{aligned}$$

The second order single-component dissipative hydrodynamics can be derived from the third moment of the distribution f^i in Israel–Stewart formalism. Its generalization to multi-component systems would be written as

$$I^{\mu\nu\alpha} = \sum_i \int \frac{g_i d^3 p}{(2\pi)^3 E_i} p_i^\mu p_i^\nu p_i^\alpha f^i, \tag{3.31}$$

and its derivative as

$$\partial_\alpha I^{\mu\nu\alpha} = \sum_i \int \frac{g_i d^3 p}{(2\pi)^3 E_i} p_i^\mu p_i^\nu p_i^\alpha \partial_\alpha f^i = Y^{\mu\nu}. \tag{3.32}$$

Here $Y^{\mu\nu}$ is a symmetric tensor to be determined later. It is non-vanishing since the third moment is not a conserving quantity unlike the second moment, which is the energy-momentum tensor. There are 10 independent components in the equations. Projection of the equations parallel and/or perpendicular to the flow u^μ yields the scalar equations $u_\mu u_\nu \partial_\alpha I^{\mu\nu\alpha} = u_\mu u_\nu Y^{\mu\nu}$ (1 component) and $\Delta_{\mu\nu} \partial_\alpha I^{\mu\nu\alpha} = \Delta_{\mu\nu} Y^{\mu\nu}$ (1 component), the vector equation $\Delta_{\rho\mu} u_\nu \partial_\alpha I^{\mu\nu\alpha} = \Delta_{\rho\mu} u_\nu Y^{\mu\nu}$ (3 components) and the tensor equation $\partial_\alpha I^{\langle\mu\nu\rangle\alpha} = Y^{\langle\mu\nu\rangle}$ (5 components).

On the other hand, the number of unknown dissipative currents is $10 + 4N$ in the systems with non-vanishing chemical potentials. In this respect, $4N$ more equations are required to fully describe the system. Since the conventional moment equations are the derivatives of the higher order moment for the energy-momentum conservation as shown in Eq. (3.32), it is natural to introduce new moment equations in multi-component systems,

$$I_J^{\mu\alpha} = \sum_i \int \frac{q_i^J g_i d^3 p}{(2\pi)^3 E_i} p_i^\mu p_i^\alpha f^i, \tag{3.33}$$

the quantum number-weighted second moments and

$$\partial_\alpha I_J^{\mu\alpha} = \sum_i \int \frac{q_i^J g_i d^3 p}{(2\pi)^3 E_i} p_i^\mu p_i^\alpha \partial_\alpha f^i = Y_J^\mu, \tag{3.34}$$

which are the derivatives of the higher order moment for the charge number conservations. One might argue that weight factors other than the conserved charge number q_i^J could have been chosen to construct the moment equations, but Eq. (3.34) would be the only vector equation which vanishes in the vanishing limit of the corresponding net charge densities. This formalism not only makes the correspondence between the number of equations and that of unknowns clear, but also allows one to construct an arbitrary number of moment equations of the form $\partial_\alpha I_J^{\mu\alpha} = Y_J^\mu$ depending on how many conserved charge currents the system in discussion possesses.

There are now 10+4N equations for 10+4N dissipative currents. The next step is to estimate $Y^{\mu\nu}$ and Y^{μ}_{J} to derive the constitutive equations in terms of the dissipative currents. The constraints are again given by the second law of thermodynamics. The definition of the entropy current (3.14) yields

$$\begin{aligned}\partial_\mu s^\mu &= -\sum_i \int \frac{g_i d^3 p_i}{(2\pi)^3 E_i} p_i^\mu \frac{\partial \phi}{\partial f^i} \partial_\mu f^i \\ &= \sum_i \int \frac{g_i d^3 p_i}{(2\pi)^3 E_i} p_i^\mu y^i \partial_\mu f^i,\end{aligned} \tag{3.35}$$

where y^i is defined in $f^i = [\exp(y^i) - \varepsilon]^{-1}$. The second order constitutive equations depend on the explicit form of the off-equilibrium distribution f^i, or equivalently, y^i. When the deviation from the local thermal equilibrium is small, one may write the off-equilibrium exponent y^i as

$$y^i = y_0^i + \delta y^i, \tag{3.36}$$

where the off-equilibrium correction δy^i is determined through Grad's moment expansion method. The conventional 14-moment method [6, 36] cannot be applied to the generic systems with N conserved charge currents because the number of macroscopic variables is then $10 + 4N$ whereas that of unknowns in the expansion remains 14. Likewise if the system has no conserved currents, i.e., $N = 0$, the system also cannot be solved, unless the concept of one conserved current is present in the system and its vanishing chemical potential limit is taken. This means that the system is implicitly assumed to be single component with binary collisions or multi-component with one conserved current in the Grad's 14-moment method. Thus one has to introduce 10+4N unknowns in the expansion as well. If one demands that (i) the distortion of the distribution be expressed in terms of the dissipative currents, (ii) the resulting constitutive equations for the dissipative currents satisfy the Onsager reciprocal relations, (iii) an arbitrary number of conserved currents be allowed in the system and (iv) the effects of a conserved current disappear in the vanishing limit of the corresponding chemical potential, then the only possible way would be to assume the expansion

$$\delta y^i = p_i^\mu \sum_J q_i^J \varepsilon_\mu^J + p_i^\mu p_i^\nu \varepsilon_{\mu\nu}, \tag{3.37}$$

where ε_μ^J and $\varepsilon_{\mu\nu}$ are macroscopic coefficients of the expansion dependent on all the components in the system. It is found later that this indeed yields the second order constitutive equations for systems with conserved charges which are reasonable compared with those of other formalisms without charges. A higher order expansion of the distribution itself can be found in Appendix E.

There are several important properties for the distortion δf^i. Firstly, the trace of the correction tensor $\varepsilon_{\mu\nu}$ in Eq. (3.37) is non-vanishing unlike that in the conventional formalism where the scalar and the traceless tensor corrections are considered. This is required for the law of increasing entropy in multi-component systems. A related discussion can be found in Appendix A. Secondly, if one considers a single component system with binary collisions, i.e., $q = 1$, then Eq. (3.37) reduces to the expansion $\delta y = p^\mu \varepsilon_\mu + p^\mu p^\nu \varepsilon_{\mu\nu}$ which is equivalent to the conventional Grad's 14-moment method in Ref. [6]. Note that this is no longer the case for the systems with chemical interactions even if the system has only one component. Thirdly, the distribution function satisfies the thermodynamic stability conditions because δe and δn_J are carefully treated in this formalism. Their vanishing limit should be taken after the constitutive equations are derived.

The unknowns in the distortion of the distribution is determined by the self-consistent conditions that the distribution should reproduce the dissipative currents when the variables are calculated in relativistic kinetic theory. Tensor decomposition of the $10 + 4N$ unknowns $\varepsilon_{\mu\nu}$ and ε^J_μ in terms of the flow is expressed as

$$\varepsilon_{\mu\nu} = \varepsilon^{\|\|} u_\mu u_\nu + \varepsilon^{\perp\mathrm{T}} \frac{\Delta_{\mu\nu}}{3} + \varepsilon^{\perp\|}_\mu u_\nu + \varepsilon^{\perp\|}_\nu u_\mu + \varepsilon^{\perp\perp}_{\mu\nu}, \tag{3.38}$$

$$\varepsilon^J_\mu = \varepsilon^{\|}_J u_\mu + \varepsilon^{\perp}_{J\mu}, \tag{3.39}$$

where the notations $\varepsilon_{\perp\mathrm{T}} = \Delta^{\mu\nu}\varepsilon_{\mu\nu}$, $\varepsilon_{\|\|} = u^\mu u^\nu \varepsilon_{\mu\nu}$, $\varepsilon^J_\| = u^\mu \varepsilon^J_\mu$, $\varepsilon^\mu_{\perp\|} = \Delta^{\mu\nu} u^\rho \varepsilon_{\nu\rho}$, $\varepsilon^{J\mu}_\perp = \Delta^{\mu\nu}\varepsilon^J_\nu$ and $\varepsilon^{\mu\nu}_{\perp\perp} = \varepsilon^{\langle\mu\nu\rangle}$ are introduced. The self-consistent conditions are then expressed separately for the scalar, the vector and the tensor parts as

$$\mathscr{J}_0 \begin{pmatrix} \varepsilon_{\perp\mathrm{T}} \\ \varepsilon_{\|\|} \\ \varepsilon^{J_1}_\| \\ \varepsilon^{J_2}_\| \\ \cdots \end{pmatrix} = \begin{pmatrix} -\Pi \\ \delta e \\ \delta n_{J_1} \\ \delta n_{J_2} \\ \cdots \end{pmatrix}, \tag{3.40}$$

$$\mathscr{J}_1 \begin{pmatrix} \varepsilon^\mu_{\perp\|} \\ \varepsilon^{J_1\mu}_\perp \\ \varepsilon^{J_2\mu}_\perp \\ \cdots \end{pmatrix} = \begin{pmatrix} W^\mu \\ V^\mu_{J_1} \\ V^\mu_{J_2} \\ \cdots \end{pmatrix}, \tag{3.41}$$

$$\mathscr{J}_2 \varepsilon^{\mu\nu}_{\perp\perp} = \pi^{\mu\nu}, \tag{3.42}$$

where the matching matrices are defined as

$$\mathscr{J}_0 = -\begin{pmatrix} \frac{5}{3}J_{42} & J_{41} & J_{31}^{j_1} & J_{31}^{j_2} & \cdots \\ J_{41} & J_{40} & J_{30}^{j_1} & J_{30}^{j_2} & \cdots \\ J_{31}^{j_1} & J_{30}^{j_1} & J_{20}^{j_1 j_1} & J_{20}^{j_1 j_2} & \cdots \\ J_{31}^{j_2} & J_{30}^{j_2} & J_{20}^{j_1 j_2} & J_{20}^{j_2 j_2} & \cdots \\ \cdots & \cdots & \cdots & \cdots & \cdots \end{pmatrix}, \tag{3.43}$$

$$\mathscr{J}_1 = -\begin{pmatrix} 2J_{41} & J_{31}^{j_1} & J_{31}^{j_2} & \cdots \\ 2J_{31}^{j_1} & J_{21}^{j_1 j_1} & J_{21}^{j_1 j_2} & \cdots \\ 2J_{31}^{j_2} & J_{21}^{j_1 j_2} & J_{21}^{j_2 j_2} & \cdots \\ \cdots & \cdots & \cdots & \cdots \end{pmatrix}, \tag{3.44}$$

$$\mathscr{J}_2 = -2J_{42}. \tag{3.45}$$

Here the moments of the distribution are defined as

$$\begin{aligned} J_{jk\cdots}^{\mu_1\cdots\mu_m} &= \sum_i \int \frac{(q_i^J q_i^K \cdots) g_i d^3p}{(2\pi)^3 E_i} f_0^i (1+\varepsilon f_0^i) p_i^{\mu_1} \cdots p_i^{\mu_m} \\ &= \sum_n \left[(\Delta^{\mu_1\mu_2} \cdots \Delta^{\mu_{2n-1}\mu_{2n}} u^{\mu_{2n+1}} \cdots u^{\mu_m}) + (\text{permutations}) \right] J_{mn}^{jk\cdots}, \end{aligned} \tag{3.46}$$

where the index $jk\cdots$ denotes the additional weight factor $q_i^J q_i^K \cdots$ in the summation over particle species i. The moments with no index indicate no weight factor in the summation. It should be noted that the moments with double weights of the same charge, e.g., $q_i^J q_i^J$ does not vanish in the limit of the corresponding vanishing chemical potential.

Equations (3.40)–(3.42) are linear equations which are easily solved as

$$\begin{aligned} \varepsilon_{\mu\nu} &= (B_\Pi \Pi + B_{\delta e}\delta e + \sum_J B_{\delta n_J}\delta n_J)\Delta_{\mu\nu} \\ &\quad + (\tilde{B}_\Pi \Pi + \tilde{B}_{\delta e}\delta e + \sum_J \tilde{B}_{\delta n_J}\delta n_J) u_\mu u_\nu \\ &\quad + 2B_W u_{(\mu} W_{\nu)} + 2\sum_J B_{V_J} u_{(\mu} V_{\nu)}^J + B_\pi \pi_{\mu\nu}, \end{aligned} \tag{3.47}$$

$$\begin{aligned} \varepsilon_\mu^J &= (D_\Pi^J \Pi + D_{\delta e}^J \delta e + \sum_K D_{\delta n_K}^J \delta n_K) u_\mu \\ &\quad + D_W^J W_\mu + \sum_K D_{V_K}^J V_\mu^K, \end{aligned} \tag{3.48}$$

where B's and D's are thermodynamic quantities which are determined by the temperature and the chemical potential only and are momentum independent. Note that they contain information of all the constituent particles in the system. The distortion factors for the dissipative currents in relativistic kinetic theory can be expressed as

$$B_\Pi = -\frac{1}{3}(\mathscr{J}_0^{-1})_{11},\ \ B_{\delta e} = \frac{1}{3}(\mathscr{J}_0^{-1})_{12},\ \ B_{\delta n_J} = \frac{1}{3}(\mathscr{J}_0^{-1})_{1,2+j}, \tag{3.49}$$

$$\tilde{B}_\Pi = -(\mathscr{J}_0^{-1})_{21},\ \ \tilde{B}_{\delta e} = (\mathscr{J}_0^{-1})_{22},\ \ \tilde{B}_{\delta n_J} = (\mathscr{J}_0^{-1})_{2,2+j}, \tag{3.50}$$

$$D_\Pi^J = -(\mathscr{J}_0^{-1})_{2+j,1},\ \ D_{\delta e}^J = (\mathscr{J}_0^{-1})_{2+j,2},\ \ D_{\delta n_K}^J = (\mathscr{J}_0^{-1})_{2+j,2+k}, \tag{3.51}$$

$$B_W = (\mathscr{J}_1^{-1})_{11},\ \ B_{V_J} = (\mathscr{J}_1^{-1})_{1,1+j}, \tag{3.52}$$

$$D_W^J = (\mathscr{J}_1^{-1})_{1+j,1},\ \ D_{V_K}^J = (\mathscr{J}_1^{-1})_{1+j,1+k}, \tag{3.53}$$

$$B_\pi = (\mathscr{J}_2^{-1})_{11}, \tag{3.54}$$

where $1 \le (k, l) \le N$. It should be noted that δy^i is expressed in Eqs. (3.47) and (3.48) as linear combinations of all the dissipative currents.

The entropy production (3.35) is now expressed as

$$\begin{aligned}\partial_\mu s^\mu &= \sum_i \int \frac{g_i d^3p}{(2\pi)^3 E_i}(y_0^i + \delta y^i) p_i^\mu \partial_\mu f^i \\ &= \sum_i \int \frac{g_i d^3p}{(2\pi)^3 E_i}\left[-\left(\sum_J \frac{q_i^J \mu_J}{T}\right)+\left(\frac{u_\nu}{T}+\sum_J q_i^J \varepsilon_\nu^J\right)p_i^\nu + \varepsilon_{\nu\rho} p_i^\nu p_i^\rho\right] p_i^\mu \partial_\mu f^i \\ &= \sum_J \varepsilon_\nu^J Y_J^\nu + \varepsilon_{\nu\rho} Y^{\nu\rho} \ge 0.\end{aligned} \tag{3.55}$$

Here energy-momentum conservation and charge number conservations are used. The tensor structures of $Y^{\nu\rho}$ and Y_J^ν in terms of the decomposition with the flow u^μ are constrained by $\varepsilon_{\nu\rho}$ and ε_ν^J for the semi-positive definite entropy production to be satisfied with arbitrary distortion of distribution. Note that again cross terms are allowed for the reason mentioned in Sect. 3.3.1. It is worth-mentioning that the finite-trace tensor correction to the distribution $\varepsilon^{\mu\nu}$ is required instead of the traceless one with the scalar correction $\tilde{\varepsilon}$ [6] for systems with multi-components and/or multiple conserved currents, because $Y_\mu^\mu \neq 0$ can be shown from the fact that the trace of Eq. (3.32) is generally non-vanishing, i.e.,

$$\partial_\alpha I_\mu^{\mu\alpha} = \sum_i m_i^2 \int \frac{g_i d^3p}{(2\pi)^3 E_i} p_i^\alpha \partial_\alpha f^i \neq 0, \tag{3.56}$$

and the existence of the moment equation for $\tilde{\varepsilon}$ violates the matching of the number of dissipative currents and that of the constitutive equations. It would give a justification to the expansion Eq. (3.37) that a non-zero trace $\varepsilon^{\mu\nu}$ should be considered. $Y^{\mu\nu}$ and Y_J^μ are then expressed as

$$\begin{pmatrix} Y_{\perp\top} \\ Y_{\| \| \|} \\ Y_{\|}^{J_1} \\ Y_{\|}^{J_2} \\ \cdots \end{pmatrix} = \mathscr{C}_0^{-1} \begin{pmatrix} \varepsilon_{\perp\top} \\ \varepsilon_{\| \| \|} \\ \varepsilon_{\|}^{J_1} \\ \varepsilon_{\|}^{J_2} \\ \cdots \end{pmatrix}, \tag{3.57}$$

$$\begin{pmatrix} Y_{\|\perp}^{\mu} \\ Y_{\perp}^{J_1\mu} \\ Y_{\perp}^{J_2\mu} \\ \cdots \end{pmatrix} = -\mathscr{C}_1^{-1} \begin{pmatrix} \varepsilon_{\|\perp}^{\mu} \\ \varepsilon_{\perp}^{J_1\mu} \\ \varepsilon_{\perp}^{J_2\mu} \\ \cdots \end{pmatrix}, \tag{3.58}$$

$$Y_{\perp\perp}^{\mu\nu} = \mathscr{C}_2^{-1} \varepsilon_{\perp\perp}^{\mu\nu}, \tag{3.59}$$

where the tensor decomposition is defined as

$$Y_{\mu\nu} = Y^{\| \| \|} u_\mu u_\nu + Y^{\perp\top} \frac{\Delta_{\mu\nu}}{3} + 2 Y_{(\mu}^{\|\perp} u_{\nu)} + Y_{\mu\nu}^{\perp\perp}, \tag{3.60}$$

$$Y_\mu^J = Y_J^{\|} u_\mu + Y_{J\mu}^{\perp}. \tag{3.61}$$

The microscopic physics of the system is integrated out in $\mathscr{C}_i$'s, the transport coefficient matrices for $\varepsilon_{\nu\rho}$ and ε_ν^J. Here $\mathscr{C}_i$'s are semi-positive definite and symmetric because of Onsager reciprocal relations. The fact that Y_J^μ and ε_μ^J approach zero in the vanishing limit of the corresponding chemical potential μ_J implies that the transport coefficients for the cross terms between $\varepsilon_{\mu\nu}$ and Y_μ^J, or equivalently, ε_μ^J and $Y_{\mu\nu}$, also vanish in the limit. The constitutive equations is uniquely determined for all the dissipative currents by solving Eqs. (3.57)–(3.59) in terms of the $10 + 4N$ variables $\varepsilon^{\mu\nu}$ and ε_J^μ and then in terms of the dissipative currents Π, δe, W^μ, $\pi^{\mu\nu}$, δn_J and V_J^μ using the explicit expressions for the distortion of distribution (3.47) and (3.48).

The constitutive equations can be obtained by estimating the derivatives $\partial_\alpha I^{\mu\nu\alpha}$ and $\partial_\alpha I_J^{\mu\alpha}$ in Eqs. (3.32) and (3.34). Since $y^i = -\sum_J \frac{q_i^J \mu_J}{T} + p_i^\mu (\frac{u_\mu}{T} + \sum_J q_i^J \varepsilon_\mu^J) + p_i^\mu p_i^\nu \varepsilon_{\mu\nu}$, their expressions up to the second order are

$$\begin{aligned} \partial_\alpha I^{\mu\nu\alpha} = {} & \sum_J J_j^{\mu\nu\alpha} \partial_\alpha \frac{\mu_J}{T} - J^{\mu\nu\alpha\beta} \partial_\alpha \frac{u_\beta}{T} - \sum_J J_j^{\mu\nu\alpha\beta} \partial_\alpha \varepsilon_\beta^J - J^{\mu\nu\alpha\beta\gamma} \partial_\alpha \varepsilon_{\beta\gamma} \\ & - \left(\sum_{J,K} K_{jk}^{\mu\nu\alpha\beta} \varepsilon_\beta^J + \sum_K K_k^{\mu\nu\alpha\beta\gamma} \varepsilon_{\beta\gamma} \right) \partial_\alpha \frac{\mu_K}{T} \\ & + \left(\sum_J K_j^{\mu\nu\alpha\beta\gamma} \varepsilon_\gamma^J + K^{\mu\nu\alpha\beta\gamma\delta} \varepsilon_{\gamma\delta} \right) \partial_\alpha \frac{u_\beta}{T}, \end{aligned} \tag{3.62}$$

$$\partial_\alpha I_J^{\mu\alpha} = \sum_K J_{jk}^{\mu\alpha} \partial_\alpha \frac{\mu_K}{T} - J_j^{\mu\alpha\beta} \partial_\alpha \frac{u_\beta}{T} - \sum_K J_{jk}^{\mu\alpha\beta} \partial_\alpha \varepsilon_\beta^K - J_j^{\mu\alpha\beta\gamma} \partial_\alpha \varepsilon_{\beta\gamma}$$

$$
- \left(\sum_{K,L} K_{jkl}^{\mu\alpha\beta} \varepsilon_\beta^K + \sum_{L} K_{jl}^{\mu\alpha\beta\gamma} \varepsilon_{\beta\gamma} \right) \partial_\alpha \frac{\mu_L}{T}
$$

$$
+ \left(\sum_{K} K_{jk}^{\mu\alpha\beta\gamma} \varepsilon_\gamma^K + K_j^{\mu\alpha\beta\gamma\delta} \varepsilon_{\gamma\delta} \right) \partial_\alpha \frac{u_\beta}{T}, \tag{3.63}
$$

where the additional moments are defined as

$$
\begin{aligned}
K_{jk\cdots}^{\mu_1\cdots\mu_m} &= \sum_i \int \frac{(q_i^J q_i^K \cdots) g_i d^3 p}{(2\pi)^3 E_i} f_0^i (1+\varepsilon f_0^i)(1+2\varepsilon f_0^i) p_i^{\mu_1} \cdots p_i^{\mu_m} \\
&= \sum_n \left[(\Delta^{\mu_1\mu_2} \cdots \Delta^{\mu_{2n-1}\mu_{2n}} u^{\mu_{2n+1}} \cdots u^{\mu_m}) + (\text{permutations}) \right] K_{mn}^{jk\cdots}.
\end{aligned} \tag{3.64}
$$

Again the index $jk\cdots$ denotes additional weight factors $q_i^J q_i^K \cdots$ in the summation over particle species. The terms in the second and the third lines of Eq. (3.62) which involve the moments K's are assumed to be small and simply omitted in the Israel–Stewart formalism [6]. It is argued in that paper that the second-order terms proportional to the Navier–Stokes thermodynamic forces, i.e., $D\frac{1}{T}$, $D\frac{\mu}{T}$, $\nabla_\mu u^\mu$, $\nabla_\mu \frac{1}{T}$, $\nabla_\mu \frac{\mu}{T}$ and $\nabla_{\langle\mu} u_{\nu\rangle}$ may not be significant while the terms proportional to acceleration Du^μ are kept in their equations. However, these should be of the same order in dissipative currents as the other second-order terms and should not be neglected to preserve consistency. Equation (3.63) corresponds to the new moment equations which do not appear in the original Israel-Stewart theory.

It is now straight-forward to derive the constitutive equations in multi-component systems with multiple conserved currents. It is obtained by combining Eqs. (3.32), (3.34), (3.40)–(3.42) and (3.57)–(3.59),

$$
\begin{pmatrix} -\Pi \\ \delta e \\ \delta n_1 \\ \delta n_2 \\ \cdots \end{pmatrix} = \mathscr{J}_0 \mathscr{C}_0 \begin{pmatrix} \frac{1}{3} \partial_\alpha I^{\perp\text{T}\alpha} \\ \partial_\alpha I^{\|\|\alpha} \\ \partial_\alpha I_{J_1}^{\|\alpha} \\ \partial_\alpha I_{J_2}^{\|\alpha} \\ \cdots \end{pmatrix}, \tag{3.65}
$$

$$
\begin{pmatrix} W^\mu \\ V_1^\mu \\ V_2^\mu \\ \cdots \end{pmatrix} = -\mathscr{J}_1 \mathscr{C}_1 \begin{pmatrix} 2\partial_\alpha I^{\perp\|\mu\alpha} \\ \partial_\alpha I_{J_1}^{\perp\mu\alpha} \\ \partial_\alpha I_{J_2}^{\perp\mu\alpha} \\ \cdots \end{pmatrix}, \tag{3.66}
$$

$$
\pi^{\mu\nu} = \mathscr{J}_2 \mathscr{C}_2 \partial_\alpha I^{\perp\perp\mu\nu\alpha}, \tag{3.67}
$$

and Eqs. (3.62) and (3.63). The second-order constitutive equations for the dissipative currents are as follows:

$$\begin{aligned}
\Pi = & -\zeta \nabla_\mu u^\mu - \zeta_{\Pi\delta e} D\frac{1}{T} + \sum_J \zeta_{\Pi\delta n_J} D\frac{\mu_J}{T} - \tau_\Pi D\Pi \\
& + \sum_J \chi^{aJ}_{\Pi\Pi} \Pi D\frac{\mu_J}{T} + \chi^b_{\Pi\Pi} \Pi D\frac{1}{T} + \chi^c_{\Pi\Pi} \Pi \nabla_\mu u^\mu \\
& + \sum_J \chi^{aJ}_{\Pi W} W_\mu \nabla^\mu \frac{\mu_J}{T} + \chi^b_{\Pi W} W_\mu \nabla^\mu \frac{1}{T} + \chi^c_{\Pi W} W_\mu D u^\mu + \chi^d_{\Pi W} \nabla^\mu W_\mu \\
& + \sum_{J,K} \chi^{aK}_{\Pi V_J} V^J_\mu \nabla^\mu \frac{\mu_K}{T} + \sum_J \chi^b_{\Pi V_J} V^J_\mu \nabla^\mu \frac{1}{T} + \sum_J \chi^c_{\Pi V_J} V^J_\mu D u^\mu + \sum_J \chi^d_{\Pi V_J} \nabla^\mu V^J_\mu \\
& + \chi_{\Pi\pi} \pi_{\mu\nu} \nabla^{\langle\mu} u^{\nu\rangle},
\end{aligned} \tag{3.68}$$

$$\begin{aligned}
W^\mu = & -\kappa_W \left(\frac{1}{T} D u^\mu + \nabla^\mu \frac{1}{T} \right) + \sum_J \kappa_{W V_J} \nabla^\mu \frac{\mu_J}{T} - \tau_W \Delta^{\mu\nu} D W_\nu \\
& + \sum_J \chi^{aJ}_{WW} W^\mu D\frac{\mu_J}{T} + \chi^b_{WW} W^\mu D\frac{1}{T} + \chi^c_{WW} W^\mu \nabla_\nu u^\nu \\
& + \chi^d_{WW} W^\nu \nabla_\nu u^\mu + \chi^e_{WW} W^\nu \nabla^\mu u_\nu \\
& - \sum_J \tau_{W V_J} \Delta^{\mu\nu} D V^J_\nu + \sum_{J,K} \chi^{aK}_{W V_J} V^\mu_J D\frac{\mu_K}{T} + \sum_J \chi^b_{W V_J} V^\mu_J D\frac{1}{T} \\
& + \sum_J \chi^c_{W V_J} V^\mu_J \nabla^\nu u_\nu + \sum_J \chi^d_{W V_J} V^\nu_J \nabla_\nu u^\mu + \sum_J \chi^e_{W V_J} V^\nu_J \nabla^\mu u_\nu \\
& + \sum_J \chi^{aJ}_{W\pi} \pi^{\mu\nu} \nabla_\nu \frac{\mu_J}{T} + \chi^b_{W\pi} \pi^{\mu\nu} \nabla_\nu \frac{1}{T} + \chi^c_{W\pi} \pi^{\mu\nu} D u_\nu + \chi^d_{W\pi} \Delta^{\mu\nu} \nabla^\rho \pi_{\nu\rho} \\
& + \sum_J \chi^{aJ}_{W\Pi} \Pi \nabla^\mu \frac{\mu_J}{T} + \chi^b_{W\Pi} \Pi \nabla^\mu \frac{1}{T} + \chi^c_{W\Pi} \Pi D u^\mu + \chi^d_{W\Pi} \nabla^\mu \Pi,
\end{aligned} \tag{3.69}$$

$$\begin{aligned}
V^\mu_J = & \kappa_{V_J} \nabla^\mu \frac{\mu_J}{T} + \sum_{K \neq J} \kappa_{V_J V_K} \nabla^\mu \frac{\mu_K}{T} - \kappa_{V_J W} \left(\frac{1}{T} D u^\mu + \nabla^\mu \frac{1}{T} \right) - \tau_{V_J} \Delta^{\mu\nu} D V^J_\nu \\
& - \sum_{K \neq J} \tau_{V_J V_K} \Delta^{\mu\nu} D V^K_\nu + \sum_{K,L} \chi^{aL}_{V_J V_K} V^\mu_K D\frac{\mu_L}{T} + \sum_K \chi^b_{V_J V_K} V^\mu_K D\frac{1}{T} \\
& + \sum_K \chi^c_{V_J V_K} V^\mu_K \nabla_\nu u^\nu + \sum_K \chi^d_{V_J V_K} V^\nu_K \nabla_\nu u^\mu + \sum_K \chi^e_{V_J V_K} V^\nu_K \nabla^\mu u_\nu \\
& - \tau_{V_J W} \Delta^{\mu\nu} D W_\nu + \sum_K \chi^a_{V_J W} W^\mu D\frac{\mu_K}{T} + \chi^b_{V_J W} W^\mu D\frac{1}{T} \\
& + \chi^c_{V_J W} W^\mu \nabla^\nu u_\nu + \chi^d_{V_J W} W^\nu \nabla_\nu u^\mu + \chi^e_{V_J W} W^\nu \nabla^\mu u_\nu
\end{aligned}$$

$$+\sum_K \chi^{aK}_{V_J\pi}\pi^{\mu\nu}\nabla_\nu\frac{\mu_K}{T}+\chi^{b}_{V_J\pi}\pi^{\mu\nu}\nabla_\nu\frac{1}{T}+\chi^{c}_{V_J\pi}\pi^{\mu\nu}Du_\nu+\chi^{d}_{V_J\pi}\Delta^{\mu\nu}\nabla^\rho\pi_{\nu\rho}$$
$$+\sum_K \chi^{aK}_{V_J\Pi}\Pi\nabla^\mu\frac{\mu_K}{T}+\chi^{b}_{V_J\Pi}\Pi\nabla^\mu\frac{1}{T}+\chi^{c}_{V_J\Pi}\Pi Du^\mu+\chi^{d}_{V_J\Pi}\nabla^\mu\Pi, \tag{3.70}$$

$$\pi^{\mu\nu}=2\eta\nabla^{\langle\mu}u^{\nu\rangle}-\tau_\pi D\pi^{\langle\mu\nu\rangle}$$
$$+\sum_J \chi^{aJ}_{\pi\pi}\pi^{\mu\nu}D\frac{\mu_J}{T}+\chi^{b}_{\pi\pi}\pi^{\mu\nu}D\frac{1}{T}+\chi^{c}_{\pi\pi}\pi^{\mu\nu}\nabla_\rho u^\rho+\chi^{d}_{\pi\pi}\pi^{\rho\langle\mu}\nabla_\rho u^{\nu\rangle}$$
$$+\sum_J \chi^{aJ}_{\pi W}W^{\langle\mu}\nabla^{\nu\rangle}\frac{\mu_J}{T}+\chi^{b}_{\pi W}W^{\langle\mu}\nabla^{\nu\rangle}\frac{1}{T}+\chi^{c}_{\pi W}W^{\langle\mu}Du^{\nu\rangle}+\chi^{d}_{\pi W}\nabla^{\langle\mu}W^{\nu\rangle}$$
$$+\sum_{J,K} \chi^{aJ}_{\pi V_J}V_J^{\langle\mu}\nabla^{\nu\rangle}\frac{\mu_K}{T}+\sum_J\chi^{b}_{\pi V_J}V_J^{\langle\mu}\nabla^{\nu\rangle}\frac{1}{T}+\sum_J\chi^{c}_{\pi V_J}V_J^{\langle\mu}Du^{\nu\rangle}+\sum_J\chi^{d}_{\pi V_J}\nabla^{\langle\mu}V_J^{\nu\rangle}$$
$$+\chi_{\pi\Pi}\Pi\nabla^{\langle\mu}u^{\nu\rangle}. \tag{3.71}$$

Here ζ's, κ's and η are the first order transport coefficients which are expressed in terms of the matrices $\mathscr{J}_i$ and the semi-positive definite matrices $\mathscr{C}_i$. As it will be discussed in Sect. 3.3.3, these transport coefficients satisfy Onsager reciprocal relations. τ_Π, τ_{V_J} and τ_π are the relaxation times and χ's are the second order transport coefficients. The stability conditions are employed at this point to take out the constitutive equations for δe and δn_J and to obtain the second order constitutive equations for Π, W^μ, V_J^μ and $\pi^{\mu\nu}$. Apparently there also is the term $\chi^e_{\pi\pi}\pi^{\rho\langle\mu}\nabla^{\nu\rangle}u_\rho$ in Eq. (3.71), but this term vanishes because orthogonality relation $\pi_{\rho\mu}u^\rho = 0$ demands $\chi^e_{\pi\pi} = 0$. The derivatives of the distortion factors D's and B's that appears in $\partial_\mu\delta f^i$ are converted to the derivatives of thermodynamic gradients using the facts that those are functions of $J^{kl\cdots}_{mn}$'s, the derivatives of which can be expressed in terms of the Navier-Stokes thermodynamic forces:

$$\begin{aligned}\partial_\mu F &= \sum_{K,L,\cdots}\sum_{m,n}\frac{\delta F}{\delta J^{kl\cdots}_{mn}}\partial_\mu J^{kl\cdots}_{mn}\\ &=\sum_J\sum_{K,L,\cdots}\sum_{m,n}\frac{\delta F}{\delta J^{kl\cdots}_{mn}}K^{jkl\cdots}_{mn}\partial_\mu\frac{\mu_J}{T}-\sum_{K,L,\cdots}\sum_{m,n}\frac{\delta F}{\delta J^{kl\cdots}_{mn}}K^{kl\cdots}_{(m+1)n}\partial_\mu\frac{1}{T}.\end{aligned} \tag{3.72}$$

Here F denotes the distortion factors B's and D's. Note that if one defines the symmetric traceless thermodynamic force as

$$\sigma^{\mu\nu}=\frac{1}{2}(\nabla^\mu u^\nu+\nabla^\nu u^\mu)-\frac{1}{3}\Delta^{\mu\nu}\nabla_\alpha u^\alpha, \tag{3.73}$$

and the vorticity as

$$\omega^{\mu\nu} = \frac{1}{2}(\nabla^\mu u^\nu - \nabla^\nu u^\mu), \tag{3.74}$$

then $\pi^{\rho\langle\mu}\nabla_\rho u^{\nu\rangle}$ in Eq. (3.71) is identically expressed as

$$\pi^{\rho\langle\mu}\nabla_\rho u^{\nu\rangle} = \pi^{\rho\langle\mu}\sigma_\rho^{\ \nu\rangle} + \pi^{\rho\langle\mu}\omega_\rho^{\ \nu\rangle} + \frac{1}{3}\pi^{\mu\nu}\nabla^\rho u_\rho, \tag{3.75}$$

which is a commonly found expression. Likewise, similar identities

$$W^\nu\nabla^\mu u_\nu = W^\nu\sigma^\mu_{\ \nu} + W^\nu\omega^\mu_{\ \nu} + \frac{1}{3}W^\mu\nabla^\nu u_\nu, \tag{3.76}$$

$$V_J^\nu\nabla^\mu u_\nu = V_J^\nu\sigma^\mu_{\ \nu} + V_J^\nu\omega^\mu_{\ \nu} + \frac{1}{3}V_J^\mu\nabla^\nu u_\nu, \tag{3.77}$$

are often used. One should be careful that $\omega^{\mu\nu}$ is an anti-symmetric tensor. In the first order limit, the constitutive equations reduce to the Navier-Stokes equations.

The derived new formalism has several major differences from the conventional method by Israel and Stewart [6]. Firstly, several second order terms which do not appear in Israel-Stewart theory are found in Eqs. (3.68)–(3.71). These terms—the second order terms composed of a dissipative current and a Navier-Stokes thermodynamic force—are obtained naturally as a result of the consistent expansion up to the second order and should also exist in the case of single component systems. Comparison with other formalisms will be presented in Sect. 5.2 where one can find that most of the second order terms reported in those literatures are also found in the newly derived equations. Secondly, linear cross terms are present in the equations. Such terms could be important from a viewpoint of non-equilibrium statistical thermodynamics. If N_J^μ is taken as a number current, the temperature gradient and acceleration in V_J^μ and the chemical gradient in W^μ represent cross-coupling phenomena known as Soret effect and Dufour effect respectively in non-equilibrium statistical mechanics [37, 38]. Likewise if the electric current is considered for the conserved current, they represent Seebeck and Peltier effects, respectively. Thirdly, there are independent second order equations for the vector dissipative currents W^μ and V_J^μ, which allow one to determine those variables in an arbitrary frame, whereas the Israel-Stewart method requires frame fixing because it yields three equations for q^μ only. Fourthly, the explicit kinetic expressions of the transport coefficients ζ's, κ's, η, τ's and χ's are different due to the existence of the new moment equations. Note that once the first order transport coefficients are given, one can estimate the second order ones including the relaxation times as they are related within the framework of relativistic kinetic theory. Finally, chemically interacting systems with multiple conserved charges can now be uniquely described.

It is also note-worthy that aside from the momentum expansion of the distribution, the constitutive equations are derived from the law of increasing entropy only.

This should be the case for a dissipative hydrodynamic formalism because it is the only thermodynamic relation which implies irreversible processes. This is in good contrast to the fact that ideal hydrodynamic equations of motion, which describe time reversible processes, are energy-momentum conservation and charge number conservations.

If one decomposes $\delta T^{\mu\nu}$ and δN_J^{μ} into each component i, one would have obtained the equations which would be much similar to the single component constitutive equations. The problems are that (a) some transport coefficients such as bulk viscosity might not be trivially separated into components, (b) in addition to the constitutive equations one needs to solve energy-momentum and charge number conservations but they do not hold for each particle species and (c) the distortion of the distribution cannot be determined without introducing additional microscopic physics, which often causes lack of generality. Such constitutive equations are not generally equivalent to the ones obtained here. Moreover, the particle species dependent equations can be applied only for the system with (quasi-)particle picture. In general, it is not the case systems in the vicinity of quark-hadron crossover.

3.3.3 Onsager Reciprocal Relations

The extended Israel–Stewart theory derived in the previous section are truncated to the first order here to find that Onsager reciprocal relations are indeed satisfied in the formalism. The entropy production is, according to Eq. (3.35), expressed up to the first order as

$$\begin{aligned}\partial_\mu s^\mu &= \sum_i \int \frac{g_i d^3 p}{(2\pi)^3 E_i} p_i^\mu \delta y^i \partial_\mu f_0^i + \mathscr{O}[\partial(\delta f^2)] \\ &= \sum_J \varepsilon_\mu^J \partial_\alpha I_{J0}^{\mu\alpha} + \varepsilon_{\mu\nu} \partial_\alpha I_0^{\mu\nu\alpha}. \end{aligned} \tag{3.78}$$

Then semi-positive definite condition of the above equation yields $10+4N$ constitutive equations. It is straight forward to derive the first order constitutive equations as

$$\begin{pmatrix} -\Pi \\ \delta e \\ \delta n_1 \\ \delta n_2 \\ \cdots \end{pmatrix} = \mathscr{A}_0 \begin{pmatrix} \frac{1}{T}\nabla_\mu u^\mu \\ D\frac{1}{T} \\ -D\frac{\mu_1}{T} \\ -D\frac{\mu_2}{T} \\ \cdots \end{pmatrix}, \tag{3.79}$$

$$\begin{pmatrix} W^\mu \\ V_1^\mu \\ V_2^\mu \\ \cdots \end{pmatrix} = -\mathscr{A}_1 \begin{pmatrix} \nabla_\mu \frac{1}{T} + \frac{1}{T} D u_\mu \\ -\nabla_\mu \frac{\mu_1}{T} \\ -\nabla_\mu \frac{\mu_2}{T} \\ \cdots \end{pmatrix}, \tag{3.80}$$

$$\pi^{\mu\nu} = \mathscr{A}_2 \frac{1}{T} \nabla^{\langle\mu} u^{\nu\rangle}. \tag{3.81}$$

The explicit forms of the transport coefficient matrices are expressed as

$$\mathscr{A}_0 = \mathscr{J}_0 \mathscr{C}_0 \mathscr{J}_0^{\mathbf{T}}, \tag{3.82}$$

$$\mathscr{A}_1 = \mathscr{J}_1 \mathscr{C}_1 \mathscr{J}_1^{\mathbf{T}}, \tag{3.83}$$

$$\mathscr{A}_2 = \mathscr{J}_2 \mathscr{C}_2 \mathscr{J}_2^{\mathbf{T}}, \tag{3.84}$$

using the moment equations (3.32)–(3.34), the matching of dissipative currents with ε's (3.40)–(3.42), the second law of thermodynamics (3.57)–(3.59) and

$$\begin{pmatrix} \frac{1}{3}\partial_\alpha I_0^{\perp T\alpha} \\ \partial_\alpha I_0^{\|\|\alpha} \\ \partial_\alpha I_{J_1 0}^{\|\alpha} \\ \partial_\alpha I_{J_2 0}^{\|\alpha} \\ \cdots \end{pmatrix} = \mathscr{J}_0^{\mathbf{T}} \begin{pmatrix} \frac{1}{T}\nabla_\mu u^\mu \\ D\frac{1}{T} \\ -D\frac{\mu_1}{T} \\ -D\frac{\mu_2}{T} \\ \cdots \end{pmatrix}, \tag{3.85}$$

$$\begin{pmatrix} 2\partial_\alpha I_0^{\perp\|\mu\alpha} \\ \partial_\alpha I_{J_1 0}^{\perp\mu\alpha} \\ \partial_\alpha I_{J_2 0}^{\perp\mu\alpha} \\ \cdots \end{pmatrix} = \mathscr{J}_1^{\mathbf{T}} \begin{pmatrix} \nabla_\mu \frac{1}{T} + \frac{1}{T} D u_\mu \\ -\nabla_\mu \frac{\mu_1}{T} \\ -\nabla_\mu \frac{\mu_2}{T} \\ \cdots \end{pmatrix}, \tag{3.86}$$

$$\partial_\alpha I_0^{\perp\perp\mu\nu\alpha} = \mathscr{J}_2^{\mathbf{T}} \frac{1}{T} \nabla^{\langle\mu} u^{\nu\rangle}. \tag{3.87}$$

Since $\mathscr{C}_i$ $(i = 0, 1, 2)$, the transport coefficient matrices for $\varepsilon^{\mu\nu}$ and ε_J^μ, are symmetric, $\mathscr{A}_i$ are also completely symmetric and Onsager reciprocal relations are satisfied. Also $\mathscr{A}_i$ are semi-positive definite because $\mathscr{C}_i$ are. This corresponds to the linear transformation of the dissipative currents and the thermodynamic forces in the entropy production discussed in Sect. 3.3.1. Note that if the distortion of the distribution other than Eq. (3.37) were employed, the reciprocal relations would not be trivially satisfied since the moment equations would no longer be uniquely constrained from the second law of thermodynamics. It is worth-mentioning while the expressions of the entropy production (3.19) and (3.78) are identical at the first order, they have second-order differences.

It should be noted here that the derivatives of the moments of equilibrium distribution in Eq. (3.78), $\partial_\alpha I_0^{\mu\nu\alpha}$ and $\partial_\alpha I_{J0}^{\mu\alpha}$, do not disappear because f_0 is the distribution for *local* thermal equilibrium, not global one. These derivatives are the sources of Navier-Stokes thermodynamic forces. The dissipative currents disappear in local thermal equilibrium because the transport coefficients vanish, not the thermodynamic forces.

3.3.4 Energy and Particle Frames

The constitutive equations for multi-component systems with multiple conserved currents (3.68)–(3.71) are frame independent and valid while the off-equilibrium deviation of the flow from ideal hydrodynamic flow is small. On the other hand, it is practically convenient to simplify the constitutive equations by choosing frames. Generality of the formalism would not be lost in frame fixing. As discussed in Sect. 3.2.1, there are two conventional ways of choosing a frame in dissipative hydrodynamics; the energy frame and the particle frame. In the energy frame, one sets the flow $u^\mu = u_E^\mu$ in the direction of the flow of energy so that no leak from fluid elements exists, i.e., $W^\mu = 0$. The constitutive equations then reduce to

$$\begin{aligned}
\Pi = &-\zeta \nabla_\mu u_E^\mu - \zeta_{\Pi\delta e} D\frac{1}{T} + \sum_J \zeta_{\Pi\delta n_J} D\frac{\mu_J}{T} - \tau_\Pi D\Pi \\
&+ \sum_J \chi_{\Pi\Pi}^{aJ} \Pi D\frac{\mu_J}{T} + \chi_{\Pi\Pi}^{b} \Pi D\frac{1}{T} + \chi_{\Pi\Pi}^{c} \Pi \nabla_\mu u_E^\mu \\
&+ \sum_{J,K} \chi_{\Pi V_J}^{aK} V_\mu^J \nabla^\mu \frac{\mu_K}{T} + \sum_J \chi_{\Pi V_J}^{b} V_\mu^J \nabla^\mu \frac{1}{T} + \sum_J \chi_{\Pi V_J}^{c} V_\mu^J D u_E^\mu + \sum_J \chi_{\Pi V_J}^{d} \nabla^\mu V_\mu^J \\
&+ \chi_{\Pi\pi} \pi_{\mu\nu} \nabla^{\langle\mu} u_E^{\nu\rangle},
\end{aligned} \tag{3.88}$$

$$\begin{aligned}
V_J^\mu = &\kappa_{V_J} \nabla^\mu \frac{\mu_J}{T} + \sum_{K\neq J} \kappa_{V_J V_K} \nabla^\mu \frac{\mu_K}{T} + \kappa_{V_J W}\left(\frac{1}{T} D u_E^\mu + \nabla^\mu \frac{1}{T}\right) - \tau_{V_J} \Delta^{\mu\nu} D V_\nu^J \\
&- \sum_{K\neq J} \tau_{V_J V_K} \Delta^{\mu\nu} D V_\nu^K + \sum_{K,L} \chi_{V_J V_K}^{aL} V_K^\mu D\frac{\mu_L}{T} + \sum_K \chi_{V_J V_K}^{b} V_K^\mu D\frac{1}{T} \\
&+ \sum_K \chi_{V_J V_K}^{c} V_K^\mu \nabla_\nu u_E^\nu + \sum_K \chi_{V_J V_K}^{d} V_K^\nu \nabla_\nu u_E^\mu + \sum_K \chi_{V_J V_K}^{e} V_K^\nu \nabla^\mu u_\nu^E \\
&+ \sum_K \chi_{V_J \pi}^{aK} \pi^{\mu\nu} \nabla_\nu \frac{\mu_K}{T} + \chi_{V_J \pi}^{b} \pi^{\mu\nu} \nabla_\nu \frac{1}{T} + \chi_{V_J \pi}^{c} \pi^{\mu\nu} D u_\nu^E + \chi_{V_J \pi}^{d} \Delta^{\mu\nu} \nabla^\rho \pi_{\nu\rho} \\
&+ \sum_K \chi_{V_J \Pi}^{aK} \Pi \nabla^\mu \frac{\mu_K}{T} + \chi_{V_J \Pi}^{b} \Pi \nabla^\mu \frac{1}{T} + \chi_{V_J \Pi}^{c} \Pi D u_E^\mu + \chi_{V_J \Pi}^{d} \nabla^\mu \Pi,
\end{aligned} \tag{3.89}$$

$$\begin{aligned}
\pi^{\mu\nu} = &2\eta \nabla^{\langle\mu} u_E^{\nu\rangle} - \tau_\pi D\pi^{\langle\mu\nu\rangle} \\
&+ \sum_J \chi_{\pi\pi}^{aJ} \pi^{\mu\nu} D\frac{\mu_J}{T} + \chi_{\pi\pi}^{b} \pi^{\mu\nu} D\frac{1}{T} + \chi_{\pi\pi}^{c} \pi^{\mu\nu} \nabla_\rho u_E^\rho + \chi_{\pi\pi}^{d} \pi^{\rho\langle\mu} \nabla_\rho u_E^{\nu\rangle} \\
&+ \sum_{J,K} \chi_{\pi V_J}^{aJ} V_J^{\langle\mu} \nabla^{\nu\rangle} \frac{\mu_K}{T} + \sum_J \chi_{\pi V_J}^{b} V_J^{\langle\mu} \nabla^{\nu\rangle} \frac{1}{T} + \sum_J \chi_{\pi V_J}^{c} V_J^{\langle\mu} D u_E^{\nu\rangle} + \sum_J \chi_{\pi V_J}^{d} \nabla^{\langle\mu} V_J^{\nu\rangle} \\
&+ \chi_{\pi\Pi} \Pi \nabla^{\langle\mu} u_E^{\nu\rangle}.
\end{aligned} \tag{3.90}$$

Note that the term $\kappa_{V_J W}(\frac{1}{T}Du_E^\mu + \nabla^\mu \frac{1}{T})$ in Eq. (3.89) does not vanish even though it contains the first order thermodynamic force for W^μ, because the constitutive equations for W^μ include the term proportional to $\kappa_{W V_J} \nabla^\mu \frac{\mu_J}{T}$ in turn.

The particle frame in single conserved current systems is defined as the frame where no leak of the charge is observed. However it is not well defined when more than one conserved current is present because in the frame where $V_J^\mu = 0$, the other currents would not vanish, i.e., $V_K^\mu \neq 0$ for $J \neq K$. One may generalize the idea and define the *average* particle frame where the sum of the charge dissipation vanishes in the case of multi-conserved current systems. The definition of the heat current q^μ in a system with N conserved currents (3.9) implies that in the average particle frame one expects $q^\mu = W^\mu$, or equivalently, $\sum_J \frac{V_J^\mu}{n_J} = 0$. The flow in this frame can be written as $u_N^\mu = \frac{1}{N}\sum_J \frac{N_J^\mu}{n_{J0}}$ which is the average of the flows in one-current particle frames. The constitutive equations in this frame have the same terms as the ones shown in Eqs. (3.68)–(3.71) because each charge current V_J^μ does not vanish.

3.3.5 Ambiguities in Phenomenological Approaches

In the following, different approaches to the derivation of second order theory are investigated to discuss their correspondences with the extended Israel-Stewart formalism. Firstly, the possibility of deriving constitutive equations from an alternative way of second-order expansion of the entropy production is explored. It is found that the equations cannot be uniquely derived this way for ambiguity in term association. As mentioned previously, the derivation of the second order constitutive equations requires the information of δf^i. If one alternatively expands Eq. (3.18) up to the second order instead of introducing the distortion of distribution from the beginning, one obtains

$$\partial_\mu s^\mu = \sum_i \int \frac{g_i d^3 p}{(2\pi)^3 E_i} p_i^\mu \bigg[\delta f^i \partial_\mu y_0^i + \delta f^i \partial_\mu \delta y^i + \frac{1}{2} f_0^i (1 + \varepsilon f_0^i)(1 + 2\varepsilon f_0^i) \delta y^{i2} \partial_\mu y_0^i \bigg], \tag{3.91}$$

which is equivalent to Eq. (3.35) at this order. Here the last term is problematic when one tries to obtain second order equations; it involves the terms with two dissipative currents coupled with one thermodynamic force, e.g., $\Pi W^\mu \nabla_\mu \frac{1}{T}$ when δy^i is estimated in the moment expansion. These terms cannot be naïvely associated with one of the dissipative currents to forcefully obtain $10 + 4N$ equations because generally the dissipative currents of different tensor structure can be found in the second order terms, e.g., $\nabla^\mu \Pi$ terms in the equation for W^μ and V^μ.

Next, a more phenomenological approach [6, 14–17], in which one expands the entropy current with respect to dissipative currents and uses the second law of thermodynamics, is investigated to see whether it yields full second order constitutive equations. If one assumes that the second order distortion of entropy current can be naïvely written as the sum of all the possible second order terms in the dissipative currents, one has

$$\begin{aligned}\delta s^{\mu}_{(2)} = \Bigg(& \alpha_0^{\Pi\Pi} \Pi^2 + \alpha_0^{\delta e \delta e} \delta e^2 + \sum_{J,K} \alpha_0^{\delta n_J \delta n_K} \delta n_J \delta n_K \\ & + \alpha_0^{\Pi \delta e} \Pi \delta e + \sum_J \alpha_0^{\Pi \delta n_J} \Pi \delta n_J + \sum_J \alpha_0^{\delta e \delta n_J} \delta e \delta n_J \\ & + \alpha_0^{WW} W^\nu W_\nu + \sum_J \alpha_0^{W V_J} W^\nu V_\nu^J + \sum_{J,K} \alpha_0^{V_J V_K} V_J^\nu V_\nu^K + \alpha_0^{\pi\pi} \pi^{\nu\rho} \pi_{\nu\rho} \Bigg) u^\mu \\ & + \alpha_1^{\Pi W} \Pi W^\mu + \sum_J \alpha_1^{\Pi V_J} \Pi V_J^\mu + \alpha_1^{\delta e W} \delta e W^\mu + \sum_J \alpha_1^{\delta e V_J} \delta e V_J^\mu \\ & + \sum_J \alpha_1^{\delta n_J W} \delta n_J W^\mu + \sum_{J,K} \alpha_1^{\delta n_J V_K} \delta n_J V_K^\mu \\ & + \alpha_1^{W\pi} W_\nu \pi^{\mu\nu} + \sum_J \alpha_1^{V_J \pi} V_\nu^J \pi^{\mu\nu}, \end{aligned} \tag{3.92}$$

where α's are undetermined coefficients. Again the derivative of the entropy current involves the terms with two dissipative currents coupled with one thermodynamic gradient, and one cannot uniquely write down the second order constitutive equations in this way. Such terms are naïvely dropped in Ref. [6].

It is note-worthy that one may introduce kinetic theory with the extended Grad's moment method to estimate the second-order correction to the entropy current (D.5). The resulting entropy current will have the same tensor structure as Eq. (3.92) does. In this case the coefficients α's are fixed in kinetic theory. When one writes down the entropy current first then takes its derivative this way, however, the terms proportional to $\Pi\pi^{\mu\nu}$ do not seem to appear in s^μ and consequently in the constitutive equations since the only possible way to construct a Lorentz vector from bulk pressure and shear stress tensor at the second order is the introduction of $\Pi\pi^{\mu\nu}u_\nu$ in the entropy current though the term is identically vanishing. This does not contradict the results obtained in Sect. 3.3.2 because there remains the ambiguity that one can add arbitrary amount of $\Pi\pi^{\mu\nu}\partial_\mu u_\nu - \pi^{\mu\nu}\Pi\partial_\mu u_\nu (= 0)$ to the entropy production because it does not contribute to the entropy production, and associate one term with the constitutive equation for the bulk pressure and the other with the ones for the shear stress tensor. This corresponds to formally keeping $\alpha_1^{\Pi\pi}\Pi\pi^{\mu\nu}u_\nu (= 0)$ in the entropy current. In other words, constitutive equations from kinetic theory in general can have the bulk-shear terms, though many conventional formalisms seem to neglect them, possibly because the amount of such terms cannot be determined.

Thus in order to determine the constitutive equations uniquely without the ambiguity of term association, it would be preferable to employ the extended Israel-Stewart formalism developed in the previous sections instead of the pure phenomenological approach or the naïve differentiation of the entropy current expansion (D.5) in kinetic theory.

3.3.6 Single-Component Systems Without Chemical Interactions

In the conventional Israel–Stewart theory, only the moment equations $\partial_\alpha I^{\mu\nu\alpha} = Y^{\mu\nu}$ (3.32) are considered. Naïvely speaking the number of equations is 10 in this approach, but if single component systems with no particle production nor annihilation are considered, the number of equations is reduced to nine because the trace of the moment equations $g_{\mu\nu}\partial_\alpha I^{\mu\nu\alpha} = g_{\mu\nu}Y^{\mu\nu}$ coincides with the number conservation $\partial_\mu N^\mu = 0$, i.e., $\partial_\alpha I_\mu^{\mu\alpha} = m^2\partial_\mu N^\mu = 0$ in kinetic theory. The two scalar equations $u_\mu u_\nu \partial_\alpha I^{\mu\nu\alpha} = u_\mu u_\nu Y^{\mu\nu}$ and $\Delta_{\mu\nu}\partial_\alpha I^{\mu\nu\alpha} = \Delta_{\mu\nu}Y^{\mu\nu}$ then become identical. On the other hand, the number of dissipative currents is 14 but can also be reduced to nine by using the stability conditions $\delta e = \delta n = 0$ and by choosing the frame to drop three independent components of W^μ or V^μ. This means that only Π, q^μ and $\pi^{\mu\nu}$ are considered. Here the heat current q^μ in single component systems is defined as

$$q^\mu = W^\mu - \frac{e_0 + P_0}{n_0} V^\mu, \tag{3.93}$$

which reduces to either W^μ or V^μ depending on whether one chooses the particle frame or the energy frame. However, the number of equations and that of dissipative currents no longer match in multi-component systems, because as mentioned earlier

$$g_{\mu\nu}\partial_\alpha I^{\mu\nu\alpha} = \sum_i m_i^2 \int \frac{g_i d^3p}{(2\pi)^3 E_i} p_i^\alpha \partial_\alpha f^i \neq 0, \tag{3.94}$$

is not a conserved current even if no particle production nor annihilation is assumed. Thus there are 10 moment equations in multi-component systems. This implies that naïve generalization of the conventional second order theory to multi-component systems does not work because only nine dissipative currents are considered there. Also if the system has more than one conserved current, the number of vector dissipative currents exceeds that of equations even if the frame is fixed. In a system with particle creations and annihilations, charge number conservations $\partial_\mu N_J^\mu = 0$ have to be considered instead of simple particle number conservation.

The apparent inconsistency between the number of equations and that of unknowns arise from the three facts. Firstly, the stability conditions $\delta e = \delta n_J = 0$ are employed prematurely and $1 + N$ unknowns are omitted in the conventional formulation. The stability conditions ensure that the system is in maximum entropy state, i.e., it is thermodynamically stable, and are independent physics from the number counting of unknown variables in kinetic theory. In this sense, the stability conditions should be introduced after the constitutive equations are derived. Secondly, the moment equations $\partial_\alpha I_J^{\mu\alpha} = Y_J^\mu$ (3.34) are not considered for the systems with conserved currents. This means that $4N$ equations are missing in the conventional formalisms. Thirdly, the heat current q^μ is considered instead of the energy current W^μ and the charge currents V_J^μ in the conventional approach, possibly due to the limitation of the number of equations. This means that 3 out of $3+3N$ vector dissipative currents are taken into account. Thus it is impossible to determine all the vector dissipative currents W^μ and V_J^μ simultaneously in an arbitrary frame within the traditional Israel-Stewart approach.

3.3.7 Correspondences with Other Formalisms

In this section, the extended Israel-Stewart formalism is compared with other frameworks of causal relativistic hydrodynamics. It should be noted that one cannot make naïve comparisons because all the other formalisms are derived in systems with single component and/or single conserved current and are sometimes dependent on the choice of frame. Therefore specific conditions such as the energy frame and the single conserved current limit are considered for clearer comparisons.

3.3.7.1 Constitutive Equations from AdS/CFT

Firstly, the results are compared with the conformal equations for $\pi^{\mu\nu}$ based on AdS/CFT correspondence [21]. The constitutive equations for the shear stress tensor from Eq. (3.71) can be expressed as, in the conformal limit $\Pi = 0$,

$$\begin{aligned} \pi^{\mu\nu} =& 2\eta\nabla^{\langle\mu}u^{\nu\rangle} - \tau_\pi D\pi^{\langle\mu\nu\rangle} + \chi_{\pi\pi}^{d}\pi^{\rho\langle\mu}\omega^{\nu\rangle}_{\ \rho} \\ &+ \chi_{\pi\pi}^{d}\pi^{\rho\langle\mu}\sigma^{\nu\rangle}_{\ \rho} + \left(\frac{1}{3}\chi_{\pi\pi}^{d} + \chi_{\pi\pi}^{c}\right)\pi^{\mu\nu}\nabla_\rho u^\rho + \chi_{\pi\pi}^{b}\pi^{\mu\nu}D\frac{1}{T}, \end{aligned} \tag{3.95}$$

when estimated in the energy frame and in the vanishing charge density limits. The identity (3.75) is used here. The first term is the Navier-Stokes term, the following two terms are the conventional second order terms and the terms in the second line are the ones that should appear in a consistent derivation as mentioned in the previous section. It is note-worthy that the term proportional to $\Pi\nabla^{\langle\mu}u^{\nu\rangle}$ is omitted in the conformal limit but its thermodynamic force is equivalent to that of $\pi^{\mu\nu}\nabla_\rho u^\rho$ term in second order theory, because both reduces to $-\zeta$(or $2\eta)\nabla^{\langle\mu}u^{\nu\rangle}\nabla_\rho u^\rho$ when the first order expressions are utilized. The former vanishes while the latter does not, because the transport coefficient ζ vanishes in the conformal limit, not the thermodynamic force itself.

On the other hand, the constitutive equations from Ref. [21] in flat space read

$$\pi^{\mu\nu} = 2\eta\nabla^{\langle\mu}u^{\nu\rangle} - \tau_\pi D\pi^{\langle\mu\nu\rangle} - \frac{d}{d-1}\tau_\pi\pi^{\mu\nu}\nabla_\rho u^\rho + \frac{\lambda_1}{\eta^2}\pi^{\rho\langle\mu}\pi^{\nu\rangle}_{\ \rho} - \frac{\lambda_2}{\eta}\pi^{\rho\langle\mu}\omega^{\nu\rangle}_{\ \rho} + \lambda_3\omega^{\rho\langle\mu}\omega^{\nu\rangle}_{\ \rho}, \tag{3.96}$$

where some of the notations are adjusted according to Eq. (3.95). Direct comparison with the extended Israel-Stewart formalism is possible because the constitutive equations of the shear stress tensor is free from the non-trivialities of multi-component systems. Note that the terms proportional to $\pi^{\mu\nu}D\frac{1}{T}$ in Eq. (3.95) can be absorbed in the $\pi^{\mu\nu}\nabla_\rho u^\rho$ term because of the ideal hydrodynamic relation (B.13). Then obvious correspondences can be found for all the terms, except for the $\omega^{\rho\langle\mu}\omega^{\nu\rangle}_{\ \rho}$ term; the terms with no dissipative currents do not appear in the Israel-Stewart-based formalism, because the distribution is expanded in terms of the dissipative currents in Grad's moment method. The AdS/CFT approach [21] yields the term because all the possible terms which are consistent with their approach are added manually in the course of derivation. If one added the vorticity-vorticity term phenomenologically in the extended Israel-Stewart formalism, that would be inconsistent in the context of the derivation in kinetic theory unless one introduces the dissipation that respond to vorticity, for example by introducing quantum anomaly [29, 39]. Discussion for non-conformal cases can be found in Ref. [25].

3.3.7.2 Constitutive Equations from Renormalization Group Method

Next the correspondences is investigated for the constitutive equations obtained in the previous sections and the ones from renormalization group approach [19]. Here a single conserved current system is considered. In the energy frame, the constitutive equations from their approach are

$$\begin{aligned}
\Pi = & -\zeta\nabla_\mu u^\mu - \tau_\Pi D\Pi \\
& + \tau_\Pi\Big[-\frac{1}{2}\frac{T\zeta}{\tau_\Pi}\partial_\mu\Big(\frac{\tau_\Pi u^\mu}{T\zeta}\Big) + \frac{1}{2}\Big(-D\frac{\mu}{T} + T\delta_\Pi^{(0)}D\frac{1}{T} + \delta_\Pi^{(1)}\nabla_\mu u^\mu\Big)\Big]\Pi \\
& + l_{\Pi V}\Big[-\nabla_\mu\frac{\mu}{T} + T\delta_{\Pi V}\Big(\nabla_\mu\frac{1}{T} + \frac{1}{T}Du_\mu\Big)\Big]V^\mu - l_{\Pi V}\nabla_\mu V^\mu \\
& + l_{\Pi\pi}\nabla_{\langle\mu}u_{\nu\rangle}\pi^{\mu\nu},
\end{aligned} \tag{3.97}$$

$$\begin{aligned}
V^\mu = & \kappa\Big(\frac{n_0 T}{e_0 + P_0}\Big)^2\nabla^\mu\frac{\mu}{T} - \tau_V\Delta^\mu_{\ \nu}DV^\nu \\
& + \tau_V\Big\{-\frac{1}{2}\frac{\kappa}{\tau_V}\Big(\frac{n_0 T}{e_0 + P_0}\Big)^2\partial_\nu\Big[\frac{\tau_V u^\nu}{\kappa}\Big(\frac{e_0 + P_0}{n_0 T}\Big)^2\Big]
\end{aligned}$$

$$+\frac{1}{2}\Big(-D\frac{\mu}{T}+T\delta_V^{(0)}D\frac{1}{T}+\frac{5}{3}\delta_V^{(1)}\nabla_\nu u^\nu\Big)\Big\}V^\mu+\tau_V\delta_V^{(1)}2\nabla^{\langle\mu}u^{\nu\rangle}V_\nu$$
$$+l_{V\pi}\Big[-\nabla_\nu\frac{\mu}{T}+T\delta_{V\pi}\Big(\nabla_\nu\frac{1}{T}+\frac{1}{T}Du_\nu\Big)\Big]\pi^{\mu\nu}-l_{V\pi}\nabla_\nu\pi^{\langle\mu\nu\rangle}$$
$$+l_{V\Pi}\Big[-\nabla^\mu\frac{\mu}{T}+T\delta_{V\Pi}\Big(\nabla^\mu\frac{1}{T}+\frac{1}{T}Du^\mu\Big)\Big]-l_{V\Pi}\nabla^\mu\Pi, \tag{3.98}$$

$$\begin{aligned}\pi^{\mu\nu}=&\,2\eta\nabla^{\langle\mu}u^{\nu\rangle}-\tau_\pi D\pi^{\langle\mu\nu\rangle}\\
&+\tau_\pi\Big[-\frac{1}{2}\frac{T\eta}{\tau_\pi}\partial_\rho\Big(\frac{\tau_\pi u^\rho}{T\eta}\Big)+\frac{1}{2}\Big(-D\frac{\mu}{T}+T\delta_\pi^{(0)}D\frac{1}{T}+\frac{7}{3}\delta_\pi^{(1)}\nabla_\rho u^\rho\Big)\Big]\pi^{\mu\nu}\\
&+\tau_\pi\delta_\pi^{(1)}4\pi^{\rho\langle\mu}\sigma^{\nu\rangle}_{\rho}\\
&+l_{\pi V}\Big[-\nabla^{\langle\mu}\frac{\mu}{T}+T\delta_{\pi V}\Big(\nabla^{\langle\mu}\frac{1}{T}+\frac{1}{T}Du^{\langle\mu}\Big)\Big]V^{\nu\rangle}-l_{\pi V}\nabla^{\langle\mu}V^{\nu\rangle}\\
&+l_{\pi\Pi}\nabla^{\langle\mu}u^{\nu\rangle}\Pi.\end{aligned} \tag{3.99}$$

Extended Israel-Stewart formalism has the vorticity terms $\omega^{\mu\nu}V_\nu$ and $\pi^{\rho\langle\mu}\omega^{\nu\rangle}_{\rho}$ whereas Eqs. (3.98) and (3.99) does not seem to have them.[2] On the other hand, the terms which involve derivatives of transport coefficients do not exist in our formalism, because they cannot be naïvely expressed with the dissipative currents, unless one assumes kinetic theory for the relaxation times over viscosities, e.g., $\frac{\tau_\Pi}{\zeta}$, and express them in terms of J_{mn}'s. It should be noted that these coefficients depend on space-time coordinates through their temperature and chemical potential dependences in the formalism. Aside from the differences due to the non-trivialities of multi-component systems, the equations have almost the same tensor structure as that in the extended Israel-Stewart formalism, possibly with different transport coefficients. The similarity in the equations is remarkable, considering that their formalism uses the technique based on renormalization group theory and is different from the kinetic theory-based formalism. It should be mentioned that their formalism has frame dependence, and the constitutive equations are different in their structures in the particle frame.

3.3.7.3 Constitutive Equations from Methods of 14 Moments

The second order constitutive equations in the original Israel-Stewart formalism [6] read

$$\Pi=-\zeta\,\partial_\mu u^\mu_E-\zeta\beta_0 D\Pi+\zeta a_0' q^\mu Du^\mu_E+\zeta\alpha_0\partial_\mu q^\mu, \tag{3.100}$$

---

2 In a recent revision, they have shown those terms are indeed present [19], which would require further investigation on the correspondence.

$$
\begin{aligned}
q^{\mu} =& \kappa T \frac{n_0 T}{e_0 + P_0} \nabla^{\mu} \frac{\mu}{T} - \kappa T \beta_1 \Delta^{\mu\nu} D q_{\nu} \\
&+ \kappa T \beta_1 \omega^{\mu\nu} q_{\nu} + \kappa T a_1 \pi^{\mu\nu} D u_{\nu}^{E} + \kappa T \alpha_1 \Delta^{\mu\nu} \partial^{\rho} \pi_{\nu\rho} \\
&+ \kappa T a_0 \Pi D u_{E}^{\mu} + \kappa T \alpha_0 \nabla^{\mu} \Pi,
\end{aligned} \tag{3.101}
$$

$$
\begin{aligned}
\pi^{\mu\nu} =& -2\eta \nabla^{\langle\mu} u_{E}^{\nu\rangle} - 2\eta\beta_2 D\pi^{\langle\mu\nu\rangle} \\
&+ 4\eta\beta_2 \pi_{\rho}^{\langle\mu} \omega^{\nu\rangle\rho} + 2\eta a_1' q^{\langle\mu} D u^{\nu\rangle} + 2\eta a_1 \partial^{\langle\mu} q^{\nu\rangle},
\end{aligned} \tag{3.102}
$$

for single component systems. Note that the metric used in the reference is opposite to the one employed in this thesis, i.e., $g^{\mu\nu} = \mathrm{diag}(-,+,+,+)$, which is the source of the negative sign before η in the first term in Eq. (3.102). Here single conserved current systems with binary collisions are considered. Compared with the extended formalism, they discard the second order terms with respect to the first order thermodynamic forces, i.e., $D\frac{1}{T}$, $D\frac{\mu}{T}$, $\nabla_{\mu} u^{\mu}$, $\nabla_{\mu}\frac{1}{T}$, $\nabla_{\mu}\frac{\mu}{T}$, and $\nabla_{\langle\mu} u_{\nu\rangle}$ stating they would be small. It is worth-mentioning, however, that the terms proportional to acceleration Du_{μ} are kept in the equations.

Several post Israel-Stewart second-order terms are found for single component systems in Refs. [14–17] by phenomenologically expressing the entropy currents in terms of the dissipative currents up to the second order. The constitutive equations are, in the particle frame,

$$
\Pi = -\zeta \partial_{\mu} u^{\mu} - \tau_{\Pi} D\Pi - \frac{1}{2}\zeta T \partial_{\mu}\left(\frac{\tau_0 u^{\mu}}{\zeta T}\right)\Pi + \tau_0 \nabla_{\mu} q^{\mu}, \tag{3.103}
$$

$$
\begin{aligned}
q^{\mu} =& \kappa(\nabla^{\mu} T - T D u^{\mu}) - \tau_q \Delta^{\mu\nu} D q_{\nu} \\
&+ \frac{1}{2}\kappa T^2 \partial_{\nu}\left(\frac{\tau u^{\nu}}{\kappa T^2}\right) q^{\mu} - \tau_1 \nabla_{\nu}\pi^{\mu\nu} - \tau_0 \nabla^{\mu}\Pi,
\end{aligned} \tag{3.104}
$$

$$
\pi^{\mu\nu} = 2\eta \nabla^{\langle\mu} u^{\nu\rangle} - \tau_{\pi} D\pi^{\langle\mu\nu\rangle} - \eta T \partial_{\lambda}\left(\frac{\tau_2 u^{\lambda}}{2\eta T}\right)\pi^{\mu\nu} + \tau_2 \nabla^{\langle\mu} q^{\nu\rangle}. \tag{3.105}
$$

The new terms here are the ones proportional to $\Pi\nabla_{\mu}u^{\mu}$, $q^{\mu}\nabla_{\nu}u^{\nu}$ and $\pi^{\mu\nu}\nabla_{\rho}u^{\rho}$. Terms which involve acceleration Du^{μ} do not appear in the second order terms. It is mentioned in the paper that when the kinetic approach is employed, vorticity terms $\tau_q \omega^{\mu\nu} q_{\nu}$ and $\tau_{\pi}\pi_{\lambda}^{\langle\mu}\omega^{\nu\rangle\lambda}$ appear in Eq. (3.104) and in Eq. (3.105), respectively. Note that due to the ambiguities of the phenomenological approaches mentioned in Sect. 3.3.5, the bulk pressure Π does not appear in the constitutive equations of the shear stress tensor $\pi^{\mu\nu}$ and vice versa.

More second order terms are reported in Ref. [27] in the framework of Israel-Stewart theory with Grad's 14-moment method. Actually when the formalism derived in this chapter is reduced for single conserved current systems and the non-trivialities of the multi-component systems are omitted, the third-moment equations from the two formalisms should become equivalent except for transport coefficients because both of them follow Israel-Stewart theory consistently. The apparent difference from our formalism comes from expansions. They expand $I^{\mu\nu\alpha}$ first and then take its derivative

$$\begin{aligned}\partial_\alpha I^{\mu\nu\alpha} =& \partial_\alpha I_0^{\mu\nu\alpha} - \partial_\alpha(J^{\mu\nu\alpha}\tilde{\varepsilon}) \\ &- \partial_\alpha(J^{\mu\nu\alpha\beta}\varepsilon_\beta) - \partial_\alpha(J^{\mu\nu\alpha\beta\gamma}\tilde{\varepsilon}_{\beta\gamma}),\end{aligned} \tag{3.106}$$

whereas the derivative of the quantity is expanded in Eq. (3.62). Here $\tilde{\varepsilon}_{\beta\gamma} = \varepsilon_{\beta\gamma} - \frac{\mathrm{Tr}(\varepsilon_{\beta\gamma})}{4} g_{\beta\gamma}$ is a traceless tensor. The coefficient $\tilde{\varepsilon}$ can be identified with $\frac{\mathrm{Tr}(\varepsilon_{\beta\gamma})}{4}$ in single component systems.

The two expansions should yield the same results up to the second order in dissipative quantities because both correspond to the expansion of the derivative of the distribution

$$\begin{aligned}\partial_\alpha f =& - f_0(1 \pm f_0)\partial_\alpha y_0 - f_0(1 \pm f_0)\partial_\alpha \delta y \\ &+ f_0(1 \pm f_0)(1 \pm 2 f_0)\delta y \partial_\alpha y_0.\end{aligned} \tag{3.107}$$

The first term on the right hand side corresponds to the Navier-Stokes limit. The second term is the source of the Israel-Stewart second order terms which includes the derivatives of the dissipative currents. Note that not all the second-order terms derived from this second term appear in the original paper by Israel and Stewart as mentioned earlier. The third term corresponds to the "new" terms that are not shown in Ref. [6]. These terms should not be neglected because they are also second order terms.

Again the energy frame is considered to compare the multi-component results with the single component ones, i.e., $q^\mu = -\frac{n_0}{e_0+P_0} V^\mu$. According to Ref. [27], the constitutive equations are

$$\begin{aligned}\Pi =& - \zeta \nabla_\mu u^\mu - \tau_\Pi D\Pi \\ &- \zeta \hat{\delta}_0 \Pi \nabla_\mu u^\mu + \tau_{\Pi q} q_\mu D u^\mu - l_{\Pi q} \partial_\mu q^\mu \\ &+ \lambda_{\Pi q} q_\mu \nabla^\mu \frac{\mu}{T} + \lambda_{\Pi\pi} \pi_{\mu\nu} \nabla^{\langle\mu} u^{\nu\rangle},\end{aligned} \tag{3.108}$$

$$\begin{aligned}q^\mu =& \kappa T \frac{n_0 T}{e_0 + P_0} \nabla^\mu \frac{\mu}{T} - \tau_q \Delta^{\mu\nu} D q_\nu \\ &- \kappa T \hat{\delta}_1 q^\mu \nabla_\nu u^\nu - \lambda_{qq} \nabla^{\langle\mu} u^{\nu\rangle} q_\nu + \tau_q \omega^{\mu\nu} q_\nu \\ &+ \lambda_{q\pi} \pi^{\mu\nu} \nabla_\nu \frac{\mu}{T} - \tau_{q\Pi} \pi^{\mu\nu} D u_\nu - l_{q\pi} \Delta^{\mu\nu} \partial^\lambda \pi_{\nu\lambda} \\ &+ \lambda_{q\Pi} \Pi \nabla^\mu \frac{\mu}{T} + \tau_{q\Pi} \Pi D u^\mu + l_{q\Pi} \nabla^\mu \Pi,\end{aligned} \tag{3.109}$$

$$\begin{aligned}\pi^{\mu\nu} =& 2\eta\nabla^{\langle\mu}u^{\nu\rangle} - \tau_\pi D\pi^{\langle\mu\nu\rangle}\\ &- 2\eta\hat{\delta}_2\pi^{\mu\nu}\nabla_\lambda u^\lambda - 2\tau_\pi\pi_\lambda{}^{\langle\mu}\sigma^{\nu\rangle\lambda} + 2\tau_\pi\pi_\lambda{}^{\langle\mu}\omega^{\nu\rangle\lambda}\\ &- 2\lambda_{\pi q}q^{\langle\mu}\nabla^{\nu\rangle}\frac{\mu}{T} + 2\tau_{\pi q}q^{\langle\mu}Du^{\nu\rangle} + 2l_{\pi q}\partial^{\langle\mu}q^{\nu\rangle}\\ &+ 2\lambda_{\pi\Pi}\Pi\nabla^{\langle\mu}u^{\nu\rangle},\end{aligned} \tag{3.110}$$

which should be identified with our constitutive equations by using the ideal hydrodynamic relations (B.13), (B.15), (B.16) and the identities (3.75) and (3.77).

A generalization of Israel-Stewart theory to a relativistic gas mixture is investigated in Ref. [10]. Their formalism is different from the extended Israel-Stewart formalism in several aspects. Firstly, the systems with no particle creation or annihilation are considered in their approach, i.e., the numbers of each particle species are conserved, while the systems with conservations based on quantum numbers are considered in this chapter. Their derivation of the equations of motion explicitly depends on the Boltzmann equation, and the dissipative currents are split for each component. On the other hand, the extended Israel-Stewart formalism aims a derivation of hydrodynamic equations and dissipative currents are not split. In the single component limit that formalism reduces to the original Israel-Stewart formalism without acceleration Du^μ and vorticity $\omega^{\mu\nu}$.

3.4 Summary

Second order constitutive equations are derived for multi-component systems with multiple conserved currents. The relaxation term, which is essential for relativistic formalism (see also Appendix F), is present in the formalism. Several new second order terms which do not appear in the original Israel–Stewart theory are obtained. It is found that naïve generalization of the conventional Israel–Stewart theory fails due to the mismatching of the number of equations and that of dissipative currents. Several non-trivialities have to be taken into account for the consistent derivation of the constitutive equations. Firstly, one must employ the stability condition after the constitutive equations are derived. This is important because premature employment of the conditions would break the correspondences between the numbers of equations and unknowns and the system could be described only when sets of assumptions are made, as one can see in the case of single component systems with one conserved current and no chemical interaction. Secondly, in systems with conserved currents, additional moment equations, which are the second moments of the distribution with conserved charges as weight factors, should be introduced. This allows one a complete description of the target systems. Thirdly, Grad's moment method for the determination of the distortion of the distribution should also be modified to match the number of equations and that of unknowns. The moment method is generalized and one finds the resulting constitutive equations satisfy Onsager reciprocal relations.

Fourthly, the law of increasing entropy requires the distortion tensor $\varepsilon_{\mu\nu}$ to have non-vanishing trace in multi-component cases.

The entropy production in relativistic kinetic theory is explicitly calculated and it is made clear that the extended Israel-Stewart theory indeed satisfies in the first order limit the Onsager reciprocal relations which demand the transport coefficient matrices to be symmetric. It is important to note that all the thermodynamic forces of the same tensor order appear in the constitutive equations. For example, the charge diffusion due to the spatial gradient in the temperature is known as Soret effect and energy dissipation due to the spatial gradient in the chemical potential as Dufour effect. Also the phenomenological approaches are found to be unsuitable for the derivation of the second order equations because they have ambiguities in associating second order terms to the constitutive equations. The second order multi-conserved current equations derived here mostly agree with the equations of other formalisms in the single conserved current limit, except for the transport coefficients and for the presence of independent constitutive equations of the vector dissipative currents W^{μ} and V_J^{μ}.

The results are very generic and should be applicable to quark-gluon or hadronic systems in high-energy heavy ion collisions as well as to early universe and cold atoms where inelastic processes of multiple conserved currents can be present.

References

1. L.D. Landau, E.M. Lifshitz, *Fluid Mechanics*, vol. 1. (Pergamon Press, New York, 1959)
2. C. Eckart, The thermodynamics of irreversible processes. 3. Relativistic theory of the simple fluid. Phys. Rev. **58**, 919 (1940)
3. W.A. Hiscock, L. Lindblom, Generic instabilities in first-order dissipative relativistic fluid theories. Phys. Rev. D **31**, 725 (1985)
4. W.A. Hiscock, L. Lindblom, Stability and causality in dissipative relativistic fluids. Ann. Phys. **151**, 466 (1983)
5. W. Israel, Nonstationary irreversible thermodynamics: a causal relativistic theory. Ann. Phys. **100**, 310 (1976)
6. W. Israel, J.M. Stewart, Transient relativistic thermodynamics and kinetic theory. Ann. Phys. **118**, 341 (1979)
7. I.-S. Liu, I. Müller, T. Ruggeri, Relativistic thermodynamics of gases. Ann. Phys. **169**, 191 (1986)
8. R.P. Geroch, L. Lindblom, Dissipative relativistic fluid theories of divergence type. Phys. Rev. D **41**, 1855 (1990)
9. B. Carter, Convective variational approach to relativistic thermodynamics of dissipative fluids. Proc. R. Soc. Lond. A **433**, 45 (1991)
10. M. Prakash, M. Prakash, R. Venugopalan, G. Welke, Nonequilibrium properties of hadronic mixtures. Phys. Rep. **227**, 321 (1993)
11. M. Grmela, H.C. Ottinger, Dynamics and thermodynamics of complex fluids. 1. Development of a general formalism. Phys. Rev. E **56**, 6620 (1997)
12. H.C. Ottinger, General projection operator formalism for the dynamics and thermodynamics of complex fluids. Phys. Rev. E **57**, 1416 (1998)
13. H.C. Öttinger, Relativistic and nonrelativistic description of fluids with anisotropic heat conduction. Phys. A **254**, 433 (1998)

14. A. Muronga, Second order dissipative fluid dynamics for ultra-relativistic nuclear collisions, Phys. Rev. Lett. **88**, 062302 (2002) [Erratum-ibid. **89**, 159901 (2002)]
15. A. Muronga, Causal theories of dissipative relativistic fluid dynamics for nuclear collisions. Phys. Rev. C **69**, 034903 (2004)
16. A. Muronga, Relativistic dynamics of non-ideal fluids: viscous and heat-conducting fluids I. General aspects and $3 + 1$ formulation for nuclear collisions. Phys. Rev. C **76**, 014909 (2007)
17. A. Muronga, Relativistic dynamics of non-ideal fluids: viscous and heat-conducting fluids II. Transport properties and microscopic description of relativistic nuclear matter. Phys. Rev. C **76**, 014910 (2007)
18. T. Koide, G.S. Denicol, Ph Mota, T. Kodama, Relativistic dissipative hydrodynamics: a minimal causal theory. Phys. Rev. C **75**, 034909 (2007)
19. T. Tsumura, T. Kunihiro, K. Ohnishi, Derivation of covariant dissipative fluid dynamics in the renormalization-group method. Phys. Lett. B **646**, 134 (2007)
20. K. Tsumura, T. Kunihiro, Second-order relativistic hydrodynamic equations for viscous systems: how does the dissipation affect the internal energy? Phys. Lett. B **690**, 255 (2010)
21. R. Baier, P. Romatschke, D.T. Son, A.O. Starinets, M.A. Stephanov, Relativistic viscous hydrodynamics, conformal invariance, and holography. J. High Energy Phys. **0804**, 100 (2008)
22. S. Bhattacharyya, V.E. Hubeny, S. Minwalla, M. Rangamani, Nonlinear fluid dynamics from gravity. J. High Energy Phys. **0802**, 045 (2008)
23. M. Natsuume, T. Okamura, Causal hydrodynamics of gauge theory plasmas from AdS/CFT duality, Phys. Rev. D **77**, 066014 (2008) [Erratum-ibid. D **78**, 089902 (2008)]
24. M. Lublinsky, E. Shuryak, Improved hydrodynamics from the AdS/CFT. Phys. Rev. D **80**, 065026 (2009)
25. P. Romatschke, Relativistic viscous fluid dynamics and non-equilibrium entropy. Class. Quantum Gravity **27**, 025006 (2010)
26. J. Peralta-Ramos, E. Calzetta, Divergence-type nonlinear conformal hydrodynamics. Phys. Rev. D **80**, 126002 (2009)
27. B. Betz, D. Henkel, D.H. Rischke, Complete second-order dissipative fluid dynamics. J. Phys. G **36**, 064029 (2009)
28. A. Monnai, T. Hirano, Relativistic dissipative hydrodynamic equations at the second order for multi-component systems with multiple conserved currents. Nucl. Phys. A **847**, 283 (2010)
29. D.T. Son, P. Surowka, Hydrodynamics with triangle anomalies. Phys. Rev. Lett. **103**, 191601 (2009)
30. P. Danielewicz, M. Gyulassy, Dissipative phenomena in quark gluon plasmas. Phys. Rev. D **31**, 53 (1985)
31. P. Kovtun, D.T. Son, A.O. Starinets, Viscosity in strongly interacting quantum field theories from black hole physics. Phys. Rev. Lett. **94**, 111601 (2005)
32. K. Tsumura, T. Kunihiro, Uniqueness of Landau-Lifshitz energy frame in relativistic dissipative hydrodynamics. arXiv:1206.3913 [physics.flu-dyn] (2012)
33. A. Monnai, T. Hirano, Effects of bulk viscosity at freezeout. Phys. Rev. C **80**, 054906 (2009)
34. L. Onsager, Reciprocal relations in irreversible processes. I. Phys. Rev. **37**, 405 (1931)
35. L. Onsager, Reciprocal relations in irreversible processes. II. Phys. Rev. **38**, 2265 (1931)
36. S.R. de Groot, W.A. van Leeuwen, Ch.G. van Weert, *Relativistic Kinetic Theory*, Chap. VII, Sect. 2. (North-Holland, Amsterdam, 1980)
37. S.R. de Groot, P. Mazur, *Non-Equilibrium Thermodynamics*, Chap. XI, Sect. 7. (North-Holland, Amsterdam, 1962)
38. D. Kondepudi, I. Prigozine, *Modern Thermodynamics: From Heat Engines to Dissipative Structures*, Sect. 16.2. (Wiley, West Sussex, 1998)
39. D.E. Kharzeev, H.-U. Yee, Anomalies and time reversal invariance in relativistic hydrodynamics: the second order and higher dimensional formulations. Phys. Rev. D **84**, 045025 (2011)

Chapter 4
Viscous Hydrodynamic Model and the Color Glass Condensate

Heavy ion collisions at LHC has marked the beginning of a new stage in the physics of the quark-gluon plasma (QGP) at high energies [1, 2]. Analyses on its first elliptic flow data suggest that the near-perfect fluid picture is still valid at LHC. It is also anticipated that the colliding nuclei are described as the color glass condensate (Sect. 2.4.1) because the energy at LHC is even higher than that at RHIC, where the calculations of the saturation model already exhibited remarkable agreements with the experimental data such as the centrality dependence of multiplicity in Au–Au and d-Au collisions. However, the latest LHC data shows that the charged particle multiplicity at mid-rapidity is $dN_{\mathrm{ch}}/d\eta_p \sim 1600$, which is considerably larger than most of the CGC predictions which indicate $\sim$1200 [2]. It should be noted here that the CGC models so far lack secondary interactions that could potentially change the rapidity distribution. Considering the models have parameters which should be constrained through comparison with the experimental data, such as λ in the saturation scale $Q_s^2(x) \sim A^{1/3}x^{-\lambda}$, it would be indispensable to build more whole picture of heavy ion collisions by integrating the viscous hydrodynamic and the CGC models for correctly interpreting the rapidity distributions.

In this chapter, viscous hydrodynamic evolution of the quark-hadron matter is estimated in the beam direction using the initial conditions from the CGC picture without assuming the boost invariant ansatz for the first time [3]. The focus is put on how ideal and viscous hydrodynamic evolution would modify the flow-rapidity dependence of the entropy that could be identified with hadronic rapidity distribution in the final stage. The full second order constitutive equations derived in the previous chapter are employed for constructing a viscous hydrodynamic model. The transverse hydrodynamic expansion is neglected in this study. We find that the viscous hydrodynamic evolution leads to visible and non-trivial deformation of the CGC initial rapidity distribution for both RHIC and LHC cases due to the interplay of the entropy production and the outward entropy flux in the systems. This could be one of the explanations for the gap between most of the CGC predictions of multiplicity at mid-rapidity and the latest LHC experimental data, because if the hydrodynamic correction lowers the multiplicity more at RHIC than at LHC, as suggested numerical estimations, the CGC parameters fixed by fitting the RHIC data without taking

A. Monnai, *Relativistic Dissipative Hydrodynamic Description of the Quark-Gluon Plasma*, Springer Theses, DOI: 10.1007/978-4-431-54798-3_4,

account of the hydrodynamic effects might under-predict the multiplicity at LHC. The result suggests that it is essential to consider a non-boost invariant viscous hydrodynamic evolution for the accurate modeling of the high-energy heavy ion collisions in the CGC based picture.

Here the term *viscous* hydrodynamics is used rather than *dissipative* hydrodynamics in the previous chapter because dissipation does not exist in the energy frame when vanishing chemical potential approximation is made. Finite-density hydrodynamic analyses will be extensively explored in Chap. 5.

4.1 Second-Order Viscous Hydrodynamics for Longitudinal Expansion

Viscous hydrodynamic model of a longitudinally expanding medium is developed in relativistic coordinates. The limit of vanishing baryon chemical potential is considered because the bulk dynamics would be described by baryon free medium at RHIC and LHC energies. Then the energy momentum decomposition (3.7) gives the definition of the thermodynamic variables in the system. The time-like (3.28) and the space-like (3.29) components of the energy-momentum conservation in a (1+1)-dimensional system are expressed as

$$De_0 = -(e_0 + P_0 + \Pi - \pi)\nabla Y_f, \tag{4.1}$$

$$(e_0 + P_0 + \Pi - \pi)DY_f = -\nabla(P_0 + \Pi - \pi), \tag{4.2}$$

where the flow rapidity Y_f is defined as $u^\mu = (\cosh Y_f, 0, 0, \sinh Y_f)$. Here the direction of expansion is set to z-axis. The time- and the space-like derivatives are

$$D = \cosh(Y_f - \eta_s)\partial_\tau + \frac{1}{\tau}\sinh(Y_f - \eta_s)\partial_{\eta_s}, \tag{4.3}$$

$$\nabla = \sinh(Y_f - \eta_s)\partial_\tau + \frac{1}{\tau}\cosh(Y_f - \eta_s)\partial_{\eta_s}, \tag{4.4}$$

which reduces to $D = \partial_\tau$ and $\nabla = \partial_{\eta_s}/\tau$ for the boost-invariant flow $Y_f = \eta_s$. It should be noted the flow is generally $Y_f \neq \eta_s$ in the current discussion. The energy dissipative current vanishes, i.e., W^μ does not exist because the energy frame is considered here for the definition of the flow.

The equations of motion for the dissipative currents are the ones derived in Chap. 3, which can be written as

$$D\Pi = \frac{1}{\tau_\Pi}\bigg(-\Pi - \zeta_{\Pi\Pi}\frac{1}{T}\nabla Y_f - \zeta_{\Pi\delta e}D\frac{1}{T} + \chi^b_{\Pi\Pi}\Pi D\frac{1}{T} + \chi^c_{\Pi\Pi}\Pi\nabla Y_f + \chi_{\Pi\pi}\pi\nabla Y_f\bigg), \tag{4.5}$$

$$D\pi = \frac{1}{\tau_\pi}\bigg(-\pi + \frac{4}{3}\eta\nabla Y_f + \chi^b_{\pi\pi}\pi D\frac{1}{T} + \chi^c_{\pi\pi}\pi\nabla Y_f + \frac{2}{3}\chi^d_{\pi\pi}\pi\nabla Y_f + \frac{2}{3}\chi_{\pi\Pi}\Pi\nabla Y_f \bigg), \quad (4.6)$$

where Π is the bulk pressure as defined in the previous chapter; π is the shear pressure defined with the shear stress tensor $\pi^{\mu\nu}$ as $\pi = \pi^{00} - \pi^{33}$. It should be noted that it is sufficient to treat π as the only independent component in the (1+1)-dimensional geometry where the system has one space dimension and one time dimension because of the orthogonality and the traceless conditions. For a reminder, η is the shear viscosity, $\zeta_{\Pi\Pi}$ and $\zeta_{\Pi\delta e}$ are the bulk viscosity and the bulk-energy cross coefficient, τ_Π and τ_π are the relaxation times for bulk and shear viscosities and $\chi^b_{\Pi\Pi}$, $\chi^c_{\Pi\Pi}$, $\chi_{\Pi\pi}$, $\chi^b_{\pi\pi}$, $\chi^c_{\pi\pi}$, $\chi^d_{\pi\pi}$ and $\chi_{\pi\Pi}$ are the second order transport coefficients. Relativistic hydrodynamic equations are generally complicated in a non-boost invariant geometry $Y_f \neq \eta_s$ because the time-like and the space-like derivatives are mixed. This can increase the numerical difficulty significantly compared with that for the transverse expansion with boost invariance.

In this study, the viscous hydrodynamic equations of motion—the energy-momentum conservation and the constitutive equations (4.5) and (4.6)—are solved in an originally developed code with the iteration method and the piecewise parabolic method [4]. The numerical difficulty for the mixing of derivatives with respect to τ and η_s is dealt by iteratively estimating the equations with an updated expansion scalar $\theta = \nabla Y_f$. The solutions are checked to converge typically in several iterative steps in numerical estimation.

4.2 The Model

4.2.1 Equation of State and Transport Coefficients

The equation of state $P_0 = P_0(e_0)$ and the transport coefficients, which reflect microscopic information of the systems, are necessary to perform hydrodynamic simulations. Here a latest result of (2+1)-flavor lattice QCD [5] is employed for the equation of state. Models are introduced for the transport coefficients since there is no single framework which determines all the transport coefficients in the equations so far. It should be noted that they are employed for the purpose of demonstration to see how visibly the entropy distribution is modified by viscosity in the hydrodynamic stage. The conjectured minimum bound $\eta/s = 1/4\pi$ from AdS/CFT correspondence [6] is used for the shear viscosity. There are two bulk viscous coefficients $\zeta_{\Pi\Pi}$ and $\zeta_{\Pi\delta e}$ as a natural consequence of formulation of dissipative hydrodynamics though so far there are few, if any, calculations on the cross coefficients. Hence an insight for these two coefficients are sought in the non-equilibrium statistical operator method [7] for the ϕ^4-theory [8]. The ratios $\zeta_{\Pi\Pi}/\eta$ and $\zeta_{\Pi\delta e}/\eta$ are parametrized as $5T/3$

and $-5T$, respectively. It is note-worthy that using energy-momentum conservation and Gibbs–Duhem relation, the two linear terms can be merged into one term at the first order in the limit of vanishing chemical potential as

$$-\zeta_{\Pi\Pi}\frac{1}{T}\nabla_\mu u^\mu - \zeta_{\Pi\delta e}D\frac{1}{T} = -\frac{\zeta_{\Pi\Pi}+c_s^2\zeta_{\Pi\delta e}}{T}\nabla_\mu u^\mu, \tag{4.7}$$

where $\zeta = (\zeta_{\Pi\Pi}+c_s^2\zeta_{\Pi\delta e})/T = 5(\frac{1}{3}-c_s^2)\eta$ corresponds to the conventional bulk viscous coefficient. See Appendix G for more details. The relaxation times and the other second order transport coefficients are estimated in kinetic theory following the discussion in the previous chapter. The second order coefficients calculated with hadronic components up to $\Delta(1232)$ are connected to those with u, d, s quarks and gluons as its components by hyperbolic factors around the (pseudo-)critical temperature using a parametrization

$$\chi = \frac{1}{2}\left(1-\tanh\frac{T-T_0}{\Delta T}\right)\chi_{\mathrm{had}} + \frac{1}{2}\left(1+\tanh\frac{T-T_0}{\Delta T}\right)\chi_{\mathrm{udsg}}, \tag{4.8}$$

where $T_0 = 0.17$ GeV and $\Delta T = 0.02$ GeV in analogy to a conventional naïve parametrization of the EoS. The relaxation times are estimated likewise. It should be emphasized again that they are meant as trial parameters for investigating dynamical responses of the hot medium qualitatively and that deriving realistic transport coefficients is not the purpose of this study.

4.2.2 Initial Conditions

Initial conditions for the hydrodynamic simulation are obtained from Monte Carlo version [9–13] of the k_T-factorization formula with unintegrated gluon distributions parametrized by Kharzeev, Levin and Nardi (MC-KLN) [14–16]. The rapidity distribution of the transverse energy distribution can be expressed as

$$\frac{dE_T}{d^2r_T dy} = \frac{4\pi N_c \alpha_s}{N_c^2-1}\int\frac{d^2p_T}{p_T}\int d^2k_T \varphi_1(x_1, k_T^2)\varphi_2(x_2,(p_T-k_T)^2), \tag{4.9}$$

where $N_c = 3$ is the number of colors, α_S is the QCD coupling, $x_{1,2} = p_T\exp(\pm y)/\sqrt{s}$ and $\varphi_{1,2}$ is the unintegrated gluon distributions, which is related to the gluon distribution as

$$xg(x) = \int^Q d^2k_T\varphi(x, k_T^2). \tag{4.10}$$

The saturation scale Q_s for a nucleus A at a transverse coordinate $r_\perp$ can be given in the model as

$$Q_{s,A}^2(x, r_\perp) = Q_{s,0}^2 \frac{T_A(r_\perp)}{T_{A,0}} \left(\frac{x_0}{x}\right)^\lambda. \tag{4.11}$$

Here $T_A(r_\perp)$ is the thickness function calculated using Woods–Saxon nuclear profile in a Glauber model-like method. $\lambda = 0.28$ is experimentally motivated. The parameters are chosen as $Q_{s,0}^2 = 2$ GeV^2, $T_{A,0} = 1.53$ fm^{-2} and $x_0 = 0.01$ [11]. The initial conditions are constructed from the 5 % most central events in the Monte-Carlo calculations. The density profile is smoothed by taking average over events.

The initial rapidity distribution of the energy density $e_0(\tau_0, \eta_s)$ is obtained from the transverse energy distribution dE_T/dy averaged over the overlapping area of the nuclei S_{area} by identifying the momentum rapidity y with the space–time rapidity η_s. The energy density is employed from the CGC model by assuming that it does not change during the short period of thermalization time. The contributions of partons with $0.1 < p_T < 3.0$ GeV/c are considered because the locally thermalized medium would be constituted mostly from the low p_T partons. No gluon production is assumed when the saturation scale is smaller than Λ_{QCD}. This leads to reduction of the gluon multiplicity at mid-rapidity by 21.5 % at RHIC and 39.9 % at LHC. The temperatures at mid-rapidity at the initial time are 419 MeV at RHIC and 490 MeV at LHC for these parameter settings. The initial flow is chosen to be the Bjorken flow $Y_f(\tau_0) = \eta_s$. Note that this boost-invariance in the flow breaks right after the hydrodynamic evolution begins.

It is generally more difficult to determine the initial conditions for the shear pressure π and the bulk pressure Π because precise dynamics in the pre-equilibrium stage is not fully known. Here they are set to vanishing because assuming finite dissipative currents at $\tau = \tau_0$ corresponds to using different initial energy-momentum tensors for ideal and viscous hydrodynamic cases and additional entropy would make comparison of the results and estimation of the viscous effects difficult.

4.3 Results

4.3.1 Rapidity Distributions at RHIC and LHC

The entropy distribution as a function of the flow rapidity dS/dY_f is investigated since it roughly corresponds to the rapidity distribution of hadrons [17]. This can be interpreted as follows; the entropy density s has the information of the number density n in the relativistic massless ideal gas limit because the ratio s/n should be dimensionless and thus temperature independent. The momentum rapidity y could be identified with the flow rapidity Y_f in a fluid element on average. Therefore dS/dY_f is the closest quantity to the rapidity distribution one can have in pure hydrodynamic analyses which are free of complicated post-hydrodynamic processes such as hadronic transport and any other model-dependent assumptions. It also allows one to compare the initial entropy distributions, which do not consist of hadrons,

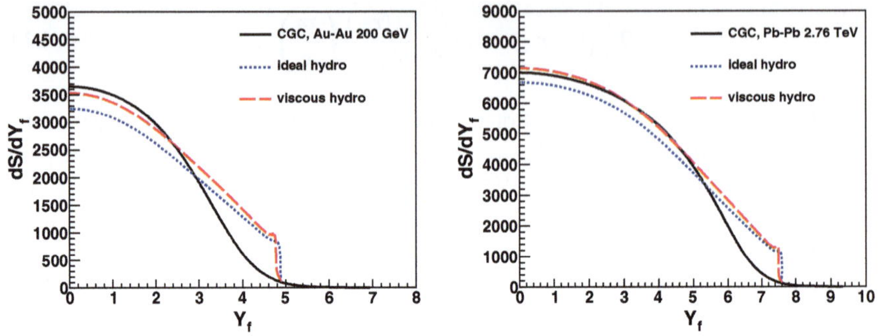

Fig. 4.1 The initial distributions dS/dY_f at $\tau = 1$ fm/c from the color glass condensate (*solid line*) and the final distributions after the ideal hydrodynamic (*dotted line*) and the shear and bulk viscous hydrodynamic (*dashed line*) evolution for (*left*) Au–Au collisions at RHIC ($\tau = 30$ fm/c) and (*right*) Pb–Pb collisions at LHC ($\tau = 50$ fm/c)

with the final hadronic rapidity distribution to estimate hydrodynamic effects. Here a freeze-out like scheme for the entropy is developed in which one constructs the rapidity distribution from the entropy flux coming out of the isothermal hypersurface of $T_f = 0.16$ GeV. Since the second order viscous correction to the entropy current s^μ should not be large, the leading order off-equilibrium modification is discussed here. A rough estimation of charged particle rapidity density can be given by

$$\frac{dN_{\text{ch}}^{\text{hydro}}}{dy} \approx \frac{2}{3} \times \frac{1}{3.6} \times \frac{dS}{dY_f}, \tag{4.12}$$

where the factor $2/3$ is for charged particles and $1/3.6$ for is for the number to entropy ratio.

Figure 4.1 shows the flow rapidity dependences of the entropy for Au–Au and Pb–Pb collisions at $\sqrt{s_{NN}} = 200$ GeV and 2.76 TeV, respectively. The typical lifetimes of the QGP here are $\tau \sim 30$ fm/c and $\tau \sim 50$ fm/c for the Au–Au and the Pb–Pb cases, respectively. Note that they are much longer than a conventional lifetime of the QGP in heavy ion collisions because the cooling process would be accelerated by transverse expansion which is neglected in the present study.

Hydrodynamic evolution visibly modifies the shapes of initial rapidity distributions from the CGC model to more flattened ones at both RHIC and LHC settings. Since the entropy is produced in non-equilibrium processes, the entropy distributions of viscous systems are larger than those of ideal systems in most of the flow rapidity region. The entropy at mid-flow rapidity is affected by the interplay between the entropy production and the outward flow which carries entropy to forward rapidity. The ideal hydrodynamic process [18] always lowers the yield of distribution at mid-flow rapidity due to the outward flow. The viscous hydrodynamic one leads to rather non-trivial results since the overall effect still lowers the distribution in the 200 GeV Au+Au collisions while it enhances the distribution in the 2.76 TeV Pb–Pb collisions.

The modification of the initial distribution is sensitive to the shape of initial distribution. The edge structure in the forward rapidity regions ($Y_f \sim$ 4–5 at RHIC and $Y_f \sim$ 7–8 at LHC) is due to the contribution of the entropy flux from the spacial hypersurface elements. The entropy carried away from mid-rapidity accumulates just before the end of the distribution to form a rather sudden fall in the forward rapidity region. It should be noted that the edge structure would be softened at freeze-out afterward by thermal smearing effects of hadronic rescattering. This, of course, does not affect the conclusion that the yield near mid-flow rapidity is affected by hydrodynamic evolution. Considering that shear and bulk viscous coefficients used in the calculation are small and close to the conjectured minimum boundaries [6, 19], the results also suggest that the effects of entropy production induced by viscosity would be important in quantitative discussion. The fact is numerically checked to remain unchanged for different choices of the initial time. Note that the initial time, along with other parameters such as viscous coefficients, are not well-known. Extraction of such parameters from experimental data would be an important and interesting task once a solid modeling of the heavy ion collisions is established. It should be noted that the numerical results here include only the longitudinal expansion and that the transverse expansion, which is missing in the present study, would become important in more quantitative analyses.

It is of particular importance that the entropy distribution $dS/dY_f \simeq dS/dy \propto dN/dy$ could be deformed from the initial condition during the hydrodynamic stage because the bare CGC gluon rapidity distribution is often directly compared with the observed charged hadron data by assuming the parton-hadron duality [14, 16, 20–24]. As discussed earlier, non-boost invariant hydrodynamic expansion can modify the initial rapidity distribution. This implies that experimentally constrained model parameters could be subject to the corrections by hydrodynamic medium interaction. If one assumes the success of the CGC models at RHIC is due to the cancellation of the entropy reduction at mid-rapidity due to convection and the entropy production due to non-equilibrium processes, additional particle production would be provided at LHC by the relatively enhanced medium effect of viscosity in comparison to that of entropy flux towards forward rapidity as suggested in Fig. 4.1. This could be a candidate explanation for the fact that most of the predictions by CGC models turned out to underestimate the multiplicity in the most central Pb–Pb collisions at LHC [2].

4.3.2 Non-boost Invariant Flow

To further quantify the deviation from boost invariance, the difference between the flow rapidity and the space–time rapidity, $Y_f - \eta_s$, are estimated. The results for Au–Au collisions with $\sqrt{s_{NN}} = 200$ GeV at $\tau = 30$ fm/c and Pb–Pb collisions with $\sqrt{s_{NN}} = 2.76$ TeV at $\tau = 50$ fm/c are shown up to the beam rapidities in Fig. 4.2. One sees in all the cases that the deviations from the boost-invariant flow are positive and become large towards larger space–time rapidity due to the acceleration by the

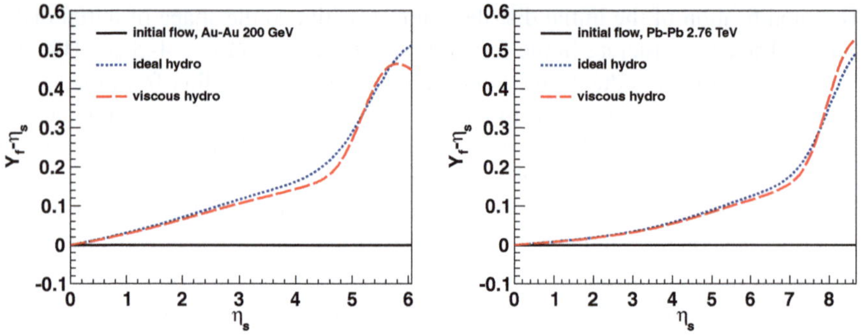

Fig. 4.2 The deviation of the flow rapidity from the space–time rapidity $Y_f - \eta_s$ at the initial time $\tau = 1$ fm/c (*solid line*) and after the ideal hydrodynamic (*dotted line*) and the shear and bulk viscous hydrodynamic (*dashed line*) evolution for (*left*) Au–Au collisions at RHIC ($\tau = 30$ fm/c) and (*right*) Pb–Pb collisions at LHC ($\tau = 50$ fm/c)

pressure gradient in the beam direction. Both Au–Au case at RHIC and Pb–Pb case at LHC exhibit the same trend, while the latter is slightly smaller near mid-rapidity possibly because the initial energy density profile is less steep at LHC. The deviations for the viscous cases are slightly smaller than those for the ideal cases for the Y_f regions in which the fluids are relatively hot, because the pressure gradients are effectively reduced in the longitudinal direction by the bulk and the shear pressures at early times in the space–time evolution. The situation, however, is reversed for later times because the entropy generated in the viscous systems enhances the pressure gradients, while the corrections from the shear and the bulk pressures themselves are already small. Due to the counter contributions at the late stage, the overall differences between the ideal and the viscous flow deviations are small at those proper times, even though the flow profiles are different in the intermediate stage. Note that while the boost-invariant initial flow is considered for the present analyses to avoid possible overestimation of the rapidity dynamics, one may also choose a non-boost invariant flow for the initial condition which would lead to enhanced convection.

4.3.3 Time Evolution of Rapidity Distribution and Flow

Finally, the time evolution of dS/dY_f and $Y_f - \eta_s$ are explicated for Pb–Pb collisions at $\sqrt{s_{NN}} = 2.76$ TeV with equal-time hypersurfaces. In Fig. 4.3 (left), the proper time dependence of the entropy distribution per the flow rapidity is shown at the times $\tau = 1$, 5, 20 and 50 fm/c. The entropy distribution does not change much its shape in the hydrodynamic evolution after $\tau \sim 20$ fm/c, suggesting that the equal-time surface at a later time is not so much different than the isothermal hypersurface except for the forward rapidity region where the medium is very dilute and can cool down quickly. The yield around $Y_f = 0$ is almost constant through-

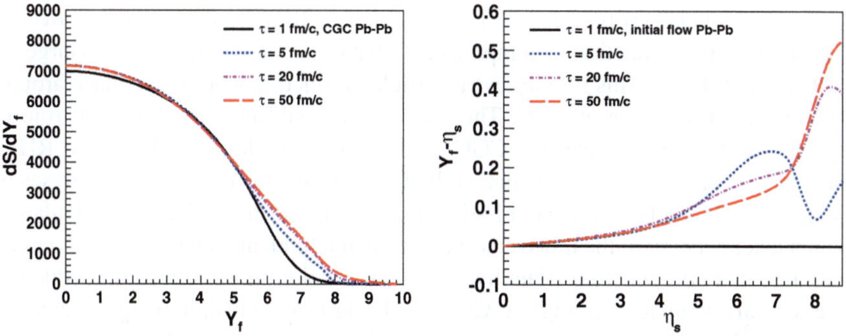

Fig. 4.3 (*Left*) The deformation of the initial entropy distribution per flow rapidity and (*right*) the deviation of the flow rapidity from the space–time rapidity $Y_f - \eta_s$ at initial time $\tau = 1$ fm/*c* (*solid line*), $\tau = 5$ fm/*c* (*dotted line*), $\tau = 20$ fm/*c* (*dash-dotted line*) and $\tau = 50$ fm/*c* (*dashed line*) in viscous hydrodynamic evolution in Pb–Pb collisions at the LHC energy

out the time evolution. It is due to a rather accidental cancellation between the entropy production and the outward entropy flux in the current parameter settings. This means a monotonous decrease would be observed at mid-flow rapidity in the case of an ideal hydrodynamic calculation. In Fig. 4.3 (right), one sees the dynamical evolution of the space–time rapidity dependence of the deviation from the boost-invariant flow. At $\tau = 5$ fm/*c*, there are an rise and a dip in the flow acceleration near $Y_f \sim$ 6–7 and 7–8, respectively, because the effective pressure $P_0 + \Pi - \pi$ can become very small in the forward flow rapidity region when the absolute values of Π and π are still large. The sudden decrease in the pressure leads to an enhancement in its gradient followed by a suppression. Note that the effect quickly disappears because the dissipative currents rapidly approach vanishing along with the time evolution. Eventually the flow rapidity distribution evolves into the one closer to the ideal hydrodynamic distribution as seen in Fig. 4.2. It is worth-mentioning that, unlike dS/dY_f, the flow rapidity profile changes in the time evolution after $\tau = 20$ fm/*c* for the current parameter sets.

4.4 Summary

To summarize, a (1+1)-dimensional second-order viscous hydrodynamic model is developed with both shear and bulk viscosities to investigate the QGP dynamics in the longitudinal direction. There is no boost-invariance in the CGC initial conditions at both RHIC and LHC energies, which leads to deformation of the entropy per the flow rapidity. This indicates that the shapes of the (pseudo-)rapidity distributions of hadrons observed in the experiments would reflect the initial gluon rapidity distributions only indirectly due to the hydrodynamic secondary interactions. This also motivates ones to readjust the model parameters which controls the rapidity dependence of entropy production. The numerical analyses indicate that the effect of viscous entropy production in comparison to that of outward entropy flux is larger at LHC than at RHIC. This implies extra particle production would take place around the mid-rapidity region at LHC. The mechanism could play an important role in explaining the gap between the CGC predictions of multiplicity based on RHIC data and the latest experimental data at LHC [2]. While the quantitative analyses of the physics should be left to a (3+1)-dimensional viscous hydrodynamic calculation which takes account of the transverse expansion, the fundamental idea should remain the same that non-boost invariant hydrodynamic evolution with entropy production is essential for understanding the CGC model in terms of high-energy heavy ion collisions. Possible future developments include introduction of the MCrcBK initial conditions [25] and the investigation of model parameter dependences.

References

1. K. Aamodt et al. [The ALICE Collaboration], Elliptic flow of charged particles in Pb–Pb collisions at 2.76 TeV. Phys. Rev. Lett. **105**, 252302 (2010)
2. K. Aamodt et al. [The ALICE Collaboration], Charged-particle multiplicity density at mid-rapidity in central Pb–Pb collisions at $\sqrt{s_{NN}} = 2.76$ TeV. Phys. Rev. Lett. **105**, 252301 (2010)
3. A. Monnai, T. Hirano, Longitudinal viscous hydrodynamic evolution for the shattered colour glass condensate. Phys. Lett. B **703**, 583 (2011)
4. P. Colella, P.R. Woodward, The piecewise parabolic method (PPM) for gas dynamical simulations. J. Comput. Phys. **54**, 174 (1984)
5. S. Borsanyi, G. Endrodi, Z. Fodor, A. Jakovac, S.D. Katz, S. Krieg, C. Ratti, K.K. Szabo, The QCD equation of state with dynamical quarks. JHEP **1011**, 077 (2010)
6. P. Kovtun, D.T. Son, A.O. Starinets, Viscosity in strongly interacting quantum field theories from black hole physics. Phys. Rev. Lett. **94**, 111601 (2005)
7. D.N. Zubarev, *Nonequilibrium Statistical Thermodynamics* (Plenum Press, New York, 1974), p. 1
8. A. Hosoya, M.A. Sakagami, M. Takao, Nonequilibrium thermodynamics in field theory: transport coefficients. Annals Phys. **154**, 229 (1984)
9. T. Hirano, Y. Nara, Hydrodynamic afterburner for the color glass condensate and the parton energy loss. Nucl. Phys. A **743**, 305 (2004)
10. T. Hirano, Y. Nara, Eccentricity fluctuation effects on elliptic flow in relativistic heavy ion collisions. Phys. Rev. C **79**, 064904 (2009)
11. A. Adil, H.-J. Drescher, A. Dumitru, A. Hayashigaki, Y. Nara, The eccentricity in heavy-ion collisions from color glass condensate initial conditions. Phys. Rev. C **74**, 044905 (2006)
12. H.-J. Drescher, Y. Nara, Effects of fluctuations on the initial eccentricity from the color glass condensate in heavy ion collisions. Phys. Rev. C **75**, 034905 (2007)
13. H.-J. Drescher, Y. Nara, Eccentricity fluctuations from the color glass condensate at RHIC and LHC. Phys. Rev. C **76**, 041903 (2007)
14. D. Kharzeev, E. Levin, Manifestations of high density QCD in the first RHIC data. Phys. Lett. B **523**, 79 (2001)
15. D. Kharzeev, E. Levin, M. Nardi, QCD saturation and deuteron nucleus collisions. Nucl. Phys. A **730**, 448 (2004) [Erratum-ibid. A **743**, 329 (2004)]
16. D. Kharzeev, E. Levin, M. Nardi, Color glass condensate at the LHC: hadron multiplicities in pp, pA and AA collisions. Nucl. Phys. A **747**, 609 (2005)
17. K. Morita, S. Muroya, C. Nonaka, T. Hirano, Comparison of space-time evolutions of hot/dense matter in $\sqrt{s_{NN}} = 17$ GeV and 130 GeV relativistic heavy ion collisions based on a hydrodynamical model. Phys. Rev. C **66**, 054904 (2002)

18. L.M. Satarov, A.V. Merdeev, I.N. Mishustin, H. Stoecker, Longitudinal fluid-dynamics for ultrarelativistic heavy-ion collisions. Phys. Rev. C **75**, 024903 (2007)
19. A. Buchel, Bulk viscosity of gauge theory plasma at strong coupling. Phys. Lett. B **663**, 286 (2008)
20. D. Kharzeev, M. Nardi, Hadron production in nuclear collisions at RHIC and high density QCD. Phys. Lett. B **507**, 121 (2001)
21. D. Kharzeev, E. Levin, M. Nardi, The onset of classical QCD dynamics in relativistic heavy ion collisions. Phys. Rev. C **71**, 054903 (2005)
22. J.L. Albacete, Particle multiplicities in lead–lead collisions at the LHC from non-linear evolution with running coupling. Phys. Rev. Lett. **99**, 262301 (2007)
23. E. Levin, A.H. Rezaeian, Hadron multiplicity in pp and AA collisions at LHC from the color glass condensate. Phys. Rev. D **82**, 054003 (2010)
24. L. McLerran, M. Praszalowicz, Saturation and scaling of multiplicity, mean p_T and p_T distributions from 200 GeV $< \sqrt{s} <$ 7 TeV. Acta Phys. Polon. B **41**, 1917 (2010)
25. J.L. ALbacete, A. Dumitru, A model for gluon production in heavy-ion collisions at the LHC with rcBK unintegrated gluon densities. arXiv:1011.5161 [hep-ph] (2010)

118. L.M. Satarov, A.V. Merdeev, I.N. Mishustin, H. Stoecker, Longitudinal fluid dynamics for ultrarelativistic heavy-ion collisions. Phys. Rev. C 75, 024903 (2007)
119. A. [illegible] Phys. Lett. B [illegible]
[illegible]

Chapter 5
Dissipative Hydrodynamic Model and Baryon Stopping

The determination of the properties of the QCD matter in wider temperature and chemical potential regions has been one of the most important goals in the hadron physics. As mentioned earlier, the quark matter in high temperature with low baryon chemical potential is experimentally accessible through high-energy heavy ion collisions, which contributes significantly to the exploration of the QCD phase diagram. At the mid-high-energy collisions in AGS, SPS and early RHIC, the baryon stopping has been used to quantify the nuclear transparency in the collisions and the kinetic energy loss for the production of the hot medium (Sect. 2.1.1). It is found that the mean rapidity losses [1, 2] at RHIC are apparently less than the simple linear scaling from AGS [3, 4] and SPS [5, 6] above around $\sqrt{s_{NN}} = 62.4\,\mathrm{GeV}$ [7, 8], which suggests that the collisions would become increasingly transparent with the collision energy. The net baryon distribution itself would be an important observable in the higher energy collisions at late RHIC and LHC as shown in a number of theoretical analyses [9–25], even though the net baryon number is often treated as irrelevant because it becomes small near mid-rapidity as the collision energy increases. It should be emphasized that the net baryon number in the initial colliding nuclei is conserved throughout the time evolution and does not vanish from the system even at highest energies. The net baryon distribution might also carry valuable information on the valence quarks in colliding nuclei and on the formation of the QGP itself in the early thermalization stage, which is not well understood yet.

The baryon number, however, is neglected in most of the modern hydrodynamic calculations with viscosity although they could be important in the context of precision physics. It is one of its great advantages that relativistic hydrodynamics can handle finite chemical potentials with ease even in off-equilibrium systems. The comparison of the experimental data to the hydrodynamical calculations of the net baryon distribution would, therefore, provide one with valuable insights for understanding the early dynamics as well as for constraining the equation of state and the transport coefficients, such as baryon diffusion coefficient, at non-vanishing baryon density. The development of a consistent dissipative hydrodynamic model at finite density would also help the efforts to explore the QCD phase diagram in the search for the critical point.

A. Monnai, *Relativistic Dissipative Hydrodynamic Description of the Quark-Gluon Plasma*, Springer Theses, DOI: 10.1007/978-4-431-54798-3_5,

In this chapter, the effects of collective flow on the net baryon distribution and the average rapidity loss for late RHIC and LHC are estimated by taking into account the interplay of shear viscosity, bulk viscosity and baryon dissipation [26]. A relativistic dissipative hydrodynamic model for the longitudinal expansion of the quark matter is developed from a generalized version of the second order theory discussed in Chap. 3. The initial energy and the initial net baryon distributions are employed from the color glass theory [27–29] and the equation of state at finite baryon density is constructed from lattice QCD results.

5.1 Finite-Density Relativistic Hydrodynamics with Dissipation

A finite-density relativistic hydrodynamic model with off-equilibrium corrections is developed for describing dynamical evolution of the quark-hadron medium in high-energy heavy ion collisions. Effect of baryon dissipation is taken into account together with those of viscosity and bulk viscosity. The equations of motion for the dissipative currents are those derived in Chap. 3 since a conserved charge is explicitly present in the target system. Cross terms among different thermodynamic forces are present both at the first and the second order in the formalism. The focus of the study is put on the longitudinal evolution of the system and the transverse dynamics is neglected as in the previous chapter. This is physically motivated by the fact that the net baryon distribution is expected to depend mainly on rapidity as the number density of net baryon is carried by valence quarks from the remnant of the colliding nuclei at forward rapidity. It is also experimentally supported that the net baryon yields do not depend on transverse geometry [30, 31] possibly because the net baryon number is homogeneous on the transverse plane unlike the pressure which explicitly depends on the geometry of the medium. It should be emphasized that the boost invariance [32] is not assumed in the longitudinal direction.

The energy-momentum tensor $T^{\mu\nu}$ and the net baryon number current N_B^μ are the conserving quantities in the system. One may introduce other conserved currents such as the net strangeness current N_s^μ, but those currents are expected to be much smaller than N_B^μ for the standard heavy ion collisions and are not considered here. In addition to the energy density, the pressure, the bulk pressure and the shear stress tensor in Eq. (3.7), tensor decomposition of the net baryon number current introduces additional thermodynamic quantities

$$N_B^\mu = (n_{B0} + \delta n_B) u^\mu + V^\mu, \tag{5.1}$$

which is an one-component case of Eq. (3.8). Here n_{B0} and V^μ denote the baryon number density and the baryon dissipation current in the local rest frame, respectively. The distortion of the baryon number density δn_{B0} equals to vanishing because of the requirement of thermodynamic stability (see Appendix D).

Linear response theory provides intuitive interpretations for the dissipative processes. The bulk pressure is a dynamical correction to the hydrostatic pressure by expansion/compression of the system without deformation, or by decrease/increase in the temperature T or the chemical potential μ_B. It is note-worthy that cancellation among the linear terms could be a reason for general smallness of the bulk pressure in many theoretical analyses since expansion and cooling are not independent phenomena in hydrodynamic systems. See Appendix G for the details. The shear stress tensor corresponds to a dynamical response to deformation without volume change. The energy and the baryon dissipation currents are the local fluxes of energy and baryon densities which dissipate away from the flow due to gradients in the temperature and the chemical potential, respectively. If one chooses the energy frame by choosing the direction of the overall local energy flux as the flow, $W^\mu = 0$ is concluded without the loss of generality. Since there is a conserved net current, one may alternatively choose the charge frame where the flow is defined with the charge flux, which leads to $V^\mu = 0$. Here the energy frame is considered for the formulation of the hydrodynamic scheme.

As seen in the previous chapter, the flow is described with the flow rapidity Y_f in (1+1)-dimensional dissipative hydrodynamics. The shear pressure $\pi = \pi^{00} - \pi^{33}$, then, is expressed as

$$\pi^{\mu\nu} = \begin{pmatrix} -\sinh^2 Y_f & 0 & 0 & -\cosh Y_f \sinh Y_f \\ 0 & \frac{1}{2} & 0 & 0 \\ 0 & 0 & \frac{1}{2} & 0 \\ -\cosh Y_f \sinh Y_f & 0 & 0 & -\cosh^2 Y_f \end{pmatrix} \pi. \tag{5.2}$$

Likewise, it follows from of the orthogonality relation $V^\mu u_\mu = 0$ that the baryon dissipation current V^μ is expressed as

$$V^\mu = (-\sinh Y_f, 0, 0, -\cosh Y_f)V, \tag{5.3}$$

where V can be called *baryon dissipation*.

Energy-momentum and net baryon number conservations are then expressed as Eq. (4.1) and

$$Dn_{B0} = -n_{B0}\nabla Y_f + \nabla V + VDY_f, \tag{5.4}$$

with the time- and the space-like derivatives D (4.3) and ∇ (4.4).

One further needs constitutive equations to determine the space-time evolution of the systems. Here the full second order dissipative hydrodynamic equations from Chap. 3, which extends the Israel-Stewart theory for the systems with particle number changing processes, are introduced. The constitutive equations in a finite-density system for the bulk pressure Π, the baryon dissipation V and the shear pressure π in the (1+1)-dimensional geometry are expressed as

$$D\Pi = \frac{1}{\tau_\Pi}\Big(-\Pi - \zeta_{\Pi\Pi}\frac{1}{T}\nabla Y_f - \zeta_{\Pi\delta e} D\frac{1}{T} + \zeta_{\Pi\delta n_B} D\frac{\mu_B}{T}$$
$$+ \chi_{\Pi\Pi}\Pi\nabla Y_f - \chi^A_{\Pi V} V\nabla\frac{\mu_B}{T} - \chi^B_{\Pi V} V\nabla\frac{1}{T}$$
$$+ \chi^C_{\Pi V} V D Y_f - \chi^D_{\Pi V}\nabla V + \chi_{\Pi\pi}\pi\nabla Y_f \Big), \tag{5.5}$$

$$DV = \frac{1}{\tau_V}\Big[-V + \kappa_{V_B}\nabla\frac{\mu_B}{T} - \kappa_{V_J W}\Big(\nabla\frac{1}{T} - \frac{1}{T} D Y_f\Big)$$
$$+ \chi_{VV} V\nabla Y_f + \chi^A_{V\pi}\pi\nabla\frac{\mu_B}{T} + \chi^B_{V\pi}\pi\nabla\frac{1}{T} - \chi^C_{V_J\pi}\pi D Y_f + \chi^D_{V\pi}\nabla\pi$$
$$+ \chi^A_{V\Pi}\Pi\nabla\frac{\mu_B}{T} + \chi^B_{V\Pi}\Pi\nabla\frac{1}{T} - \chi^C_{V\Pi}\Pi D Y_f + \chi^D_{V\Pi}\nabla\Pi\Big], \tag{5.6}$$

$$D\pi = \frac{1}{\tau_\pi}\Big(-\pi + \frac{4}{3}\eta\nabla Y_f$$
$$+ \chi_{\pi\pi}\pi\nabla Y_f + \chi_{\pi\Pi}\Pi\nabla Y_f - \chi^A_{\pi V} V\nabla\frac{\mu_B}{T}$$
$$- \chi^B_{\pi V} V\nabla\frac{1}{T} + \chi^C_{\pi V} V D Y_f - \chi^D_{\pi V}\nabla V\Big), \tag{5.7}$$

where again η is the shear viscosity, $\zeta_{\Pi\Pi}$, $\zeta_{\Pi\delta e}$ and $\zeta_{\Pi\delta n_B}$ are the bulk viscosities, κ_V is the baryon charge conductivity and κ_{VW} is the baryon-heat cross conductivity. It should be noted that the linear cross terms, which satisfy Onsager reciprocal relations [33, 34], are explicitly present in the hydrodynamic framework. The (diagonal) baryon charge conductivity is related with the baryon diffusion coefficient as

$$D_B = \left(\frac{\partial\mu_B}{\partial n_{B0}}\right)_T \frac{\kappa_V}{T}. \tag{5.8}$$

The cross conductivity causes the Soret effect, which is the chemical diffusion induced by the thermal gradient and flow acceleration. It is related to the thermo-diffusion coefficient as

$$D_T = \left(\frac{\partial\mu_B}{\partial T}\right)_{P_0} \frac{\kappa_V}{n_{B0}} + \frac{\kappa_{VW} - \mu_B\kappa_V}{T}. \tag{5.9}$$

The ratio $k_T = D_T/D_B$ is called Soret coefficient. The cross coefficients can either be positive or negative while the semi-positive definite condition of the transport coefficient matrix $\kappa^2_{VW} \leq \kappa_V\kappa_W$ is satisfied. Here κ_W is the thermal conductivity which does not explicitly appear in the formalism because of the frame choice. τ_Π, τ_π and τ_V are the relaxation times, $\chi_{\Pi\Pi}$, $\chi_{\pi\pi}$ and χ_{VV} are the second order self-coupling coefficients, and $\chi_{\Pi\pi}$, $\chi_{\pi V}$ and $\chi_{V\Pi}$ are the second order cross coefficients. Some of the second-order terms have different indices from the ones introduced in Chap. 3 because they are combined using the conservation laws and the Gibbs-Duhem relation with the truncation to the second order. The generality of the equations is not lost in the prescription. It should be noted that the couplings among different dissipative quantities at the second order can in principle have non-trivial effects because quantitative hierarchies can be present among the dissipative quantities as is indeed the case for quark-hadron systems.

The equations have many time-like derivatives, potentially increasing the numerical difficulties. Here they are solved with an advanced version of the multiple iteration algorithm with the piece-wise parabolic method [35] in Chap. 4.

5.2 The Model

The relativistic hydrodynamic model describes macroscopic motion of a fluid with the conservation laws and the constitutive equations. The QCD equation of state and the QCD transport coefficients must be given as parameter input to perform hydrodynamic calculations for the quark-hadron matter. Since the hydrodynamic description works in the intermediate stage of the heavy ion collisions around $\tau \sim$ 1–10 fm/c, one further needs to introduce initial conditions and freeze-out to link hydrodynamic analyses with experimental observables. Here the color glass theory is employed to estimate the initial conditions for the energy density and the net baryon density profiles. The hydrodynamic flow is converted to particles at freeze-out with Cooper–Fyre prescription explained later.

5.2.1 Equation of State and Transport Coefficients at Finite Baryon Density

The equation of state (EoS) and the transport coefficients represent static and dynamical responses of a thermodynamic system that come from microscopic properties of the medium, and are necessary input for a hydrodynamic model. Obtaining these quantities from first principle calculations is generally a very non-trivial issue especially for systems with finite baryon chemical potential due to the fermion-sign problem in the lattice QCD formulations. Here a finite-density EoS is constructed with the Taylor expansion method around the vanishing chemical potential up to the second order,

$$\frac{P(T,\mu_B)}{T^4} = \frac{P(T,0)}{T^4} + \frac{\chi_B^{(2)}(T,0)}{2}\left(\frac{\mu_B}{T}\right)^2 + \mathcal{O}\left(\frac{\mu_B}{T}\right)^4, \tag{5.10}$$

where $\chi_B^{(2)}$ is the quadratic fluctuation of the baryon number. Latest (2+1)-flavor lattice QCD results of the continuum extrapolations are employed for the EoS at vanishing chemical potentials [36] and for the quadratic baryon fluctuation [37]. It should be noted that the chemical potential to temperature ratio μ_B/T has to be small for the expansion to be formally valid. This is well motivated for the high-energy heavy ion collisions where the net baryon density is relatively small.

The transport coefficients of the hot matter, including the finite-density ones, are more difficult to obtain in the first principle calculations. This makes constraining the coefficients from experimental data in a dissipative hydrodynamic modeling one of the goals of heavy ion physics. Here model coefficients are introduced for demonstrative purposes. The shear viscous coefficient is again employed from the AdS/CFT minimum boundary $\eta/s = 1/4\pi$ [38]. Note that the quantity is implicitly subject to finite-density corrections as $s = (e_0 + P_0 - \mu_B n_{B0})/T$. Following the discussion in Appendix G, the three bulk viscous coefficients are combined to *effective* bulk viscous coefficient $\zeta = 5(\frac{1}{3} - c_s^2)\eta$ at the first order. Here the squared sound velocity is $c_s^2 = (\partial P_0/\partial e_0)_{s/n_{B0}}$. The coefficient phenomenologically exhibits a peak structure around the crossover temperature T_c due to reduction in the sound velocity.

The baryon charge conductivity and the baryon-heat cross coefficient characterize the non-equilibrium processes at finite baryon density. Using the charge conductivity $D_B = 1/2\pi T$ from an AdS/CFT framework [39], the former is estimated as

$$\kappa_V = \frac{c_V}{2\pi}\left(\frac{\partial \mu_B}{\partial n_B}\right)_T^{-1} = c_V \frac{\chi_B^{(2)} T^2}{2\pi}, \tag{5.11}$$

where the Taylor expansion of the EoS (5.10) is utilized. c_V is a dimensionless constant introduced for parametrization. Here $c_V = 1$ is considered unless specified otherwise. The cross coefficient is parametrized from the dimensional analyses, the matter-antimatter symmetry $V^\mu(\mu_B) = -V^\mu(-\mu_B)$ and the implication from the semi-positive definiteness of the transport coefficient matrix as $\kappa_{VW} = c_{VW}[n_{B0}T/(e_0 + P_0)]\sqrt{\kappa_V \kappa_W}$ where c_{VW} is a parameter constant. Here the thermal conductivity is chosen as $\kappa_W = 5\eta T$ [40]. The semi-positive definite condition for the conductivities in this case is explicitly expressed as $c_{VW}^2[n_{B0}T/(e_0 + P_0)]^2 \leq 1$. Note that the charge conductivity in this model remains finite in the limit of vanishing chemical potential, but does not induce the charge dissipation current out of global chemical equilibrium because the cross coefficients vanish in the limit and the charge conductivity forms an isolated partial matrix in the full transport coefficient matrix. This property is due to the matter-antimatter symmetry. The lack of cross coefficients at the vanishing limit of the corresponding chemical potentials is important because it ensures that hidden conserving quantities—strangeness or yet unknown charge—do not affect the physics. The temperature dependences of the dimensionless ratios of the linear transport coefficients η/s, ζ/s, $\kappa_V T/s$ and κ_{VW}/s with $c_V = 1$ and $c_{WV} = 5$ for a constant chemical potential $\mu_B = 0.05$ GeV are plotted in Fig. 5.1. One can see that the baryon-related transport coefficients are smaller than the others in the current parameter settings.

The relaxation times are also estimated from an AdS/CFT framework as $\tau_\pi = (2 - \ln 2)/2\pi T$, $\tau_\Pi = 18 - (9 \ln 3 - \sqrt{3}\pi)/24\pi T$ and $\tau_V = \ln 2/2\pi T$ [39]. The other

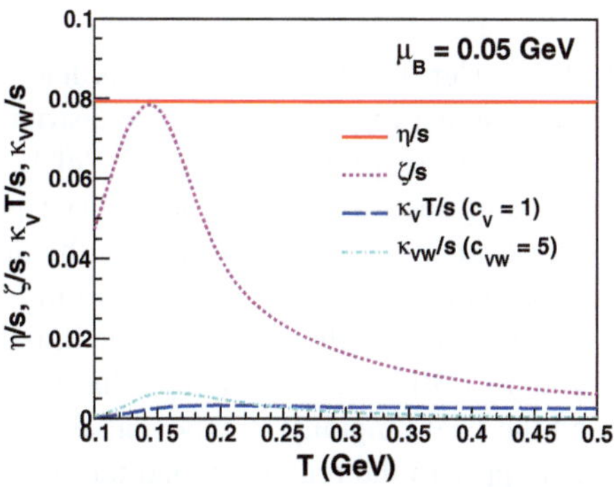

Fig. 5.1 The dimensionless ratios of the shear viscosity (*solid line*), the effective bulk viscosity (*dotted line*), the baryon charge conductivity multiplied by temperature (*dashed line*) and the baryon-heat cross conductivity (*dash-dotted line*) to the entropy density at $\mu_B = 0.05$ GeV. The temperature region relevant to the numerical simulation $0.1 \leq T \leq 0.5$ GeV is shown

second order transport coefficients are parametrized as $\chi_{AB} = \tilde{c}_{AB}[n_{B0}T/(e_0 + P_0)]^n T^m \tau_A$ where A and B denote the types of dissipative currents and $\tilde{c}_{AB}$ is a dimensionless constant. $n = 1$ is employed for the baryon-non-baryon cross coefficients and $n = 0$ for the others, following the matter-antimatter symmetry arguments earlier. m is the dimension parameter chosen so that the dimensions of the terms in each constitutive equation are matched. For the most part in this study, however, $\tilde{c}_{AB} = 0$ is employed to investigate the qualitative nature of the dissipative processes. Note that, in the present analyses, the focus is on the net baryon density and the interplay of different dissipative currents, and it is beyond the scope of the current analyses to obtain the precise values of the transport coefficients.

5.2.2 *Initial Conditions*

The initial condition for the energy density is constructed from the color glass condensate model. Since the gluons are dominant against the valence quarks in high-energy heavy ion collisions, the initial energy density of the hot matter is estimated from the gluon distribution by assuming the profile of the energy distribution is not significantly modified during the early thermalization stage. The Nara Monte-Carlo adaptation [41–45] of Kharzeev-Levin-Nardi model (MC-KLN) [46] is employed for the estimation as in Chap. 4.

The initial condition for the net baryon density is also constructed from the color glass picture by assuming it is proportional to the valence quark parton distribution function [22]. The net baryon distribution for a nucleus reads

$$\frac{dN_{B-\bar{B}}}{dy} = \frac{C}{(2\pi)^2} \int \frac{d^2 p_T}{p_T^2} x_1 q_v(x_1) \varphi(x_2, p_T), \tag{5.12}$$

where q_v is the valence quark distribution, φ is the unintegrated gluon distribution and $x_{1,2} = p_T \exp(\pm y)/\sqrt{s}$. Here the normalization C is determined so that the integrated $N_{B-\bar{B}}$ matches the number of participants. The distributions are determined with the same settings as the no-fragmentation case in Ref. [22] but with $\lambda = 0.28$ and the next-to-next-to-leading order fit results of the valence quark distribution [47]. Again the distribution is assumed not to be modified in the pre-equilibrium evolution. Note that the net baryon distribution is sensitive to the parameters, and here they are chosen to yield much steeper distribution at the initial stage so the distribution after the hydrodynamic evolution roughly reproduces experimental data. The tail contribution beyond the beam rapidity is exponentially cut off since hydrodynamic description do not apply to the region beyond the freeze-out and the Taylor expansion-based EoS is not expected to work for the dense and cold matter. This leads to the reduction of the total baryon number of participant nucleons by 8.7 % at RHIC and 0.4 % at LHC for the most central 0–5% events.

The initial conditions for the dissipative currents are not well known and this is also a quite interesting issue by itself as they are related with the magnitude of thermalization in the pre-equilibrium stage. Here they are simply chosen to be non-existent at the initial time, i.e., $\Pi(\tau_0, \eta_s) = V(\tau_0, \eta_s) = \pi(\tau_0, \eta_s) = 0$ for a clear view of non-equilibrium effects on the fluids and also for avoiding possible overestimation of viscous and baryon dissipative effects from ambiguity in the early stage. Here τ_0 is the initial time set to $\tau_0 = 1$ fm/c.

5.2.3 Freeze-Out

As the expanding system cools down with the time evolution and becomes dilute enough, the hydrodynamic simulation has to be stopped and the flow field needs to be converted into particles. One conventionally employs the Cooper–Frye formula [48] at a freeze-out hypersurface Σ. The rapidity distribution reads

$$\frac{dN_{B-\bar{B}}}{dy} = \sum_i \frac{b_i g_i}{(2\pi)^3} \int d^2 p_T \int_\Sigma p_i^\mu d\sigma_\mu f_i, \tag{5.13}$$

where b_i is the baryon number, i.e., $b_i = +1$ for baryons, -1 for anti-baryons and 0 for mesons. g_i is the degeneracy, $d\sigma_\mu$ is the freeze-out hypersurface element and f_i is the phase-space distribution. Σ is taken as an isothermal surface because the chemical potential dependence of the boundary is sufficiently small for the high-energy collisions. This gives rise to the concept of the freeze-out temperature T_f. f_i can be separated into the equilibrium distribution f_i^0 and the distortion of distribution δf_i, where

$$f_i^0 = \left[\exp\left(\frac{p_\mu^i u^\mu - b_i \mu_B}{T} \right) - \varepsilon_i \right]^{-1}. \tag{5.14}$$

Here ε_i denotes the quantum statistics as $\varepsilon_i = +1$ for fermions and -1 for bosons. The effects of the off-equilibrium distribution δf_i is not treated here because its correction on the p_T-integrated net baryon rapidity distribution would be small when the off-equilibrium expansion is applicable, since the stability condition for the net baryon density requires $\delta n_B = 0$ and $|\delta N_B^\mu|/|N_{B0}^\mu| = V/n_{B0}$ would be generally very small. In general it could be important, especially for transverse particle spectra, as discussed in Appendix A for shear and bulk viscosities.

It should be emphasized that the focus of this chapter is to estimate the finite-density hydrodynamic effects with non-equilibrium processes on the hot medium because it would have dominant effects on the net baryon distribution. The hadronic cascade at the later stage is not considered because the modification on the net baryon distribution during the hadronic stage would be small as the baryon number does not change in the hadronic decay and diffusion process in a hadronic gas is expected to be slow [14, 49].

The information on the chemical freeze-out is implicitly contained in the lattice QCD EoS in the hadronic phase since the hydrodynamic flow does not specify its contents. If the EoS for each hadronic component was known, one could in principle incorporate the chemical freeze-out explicitly by introducing conservation laws for the hadrons instead of the one for the baryon charge [50].

5.3 Results

The most central 0–5 % events are considered for constructing the initial conditions. The mean numbers of participants in the model are 357 for RHIC and 385 for LHC. The initial temperatures and the chemical potentials at mid-rapidity are $T = 419$ MeV and $\mu_B = 20.6$ MeV for RHIC and $T = 490$ MeV and $\mu_B = 4.3$ MeV for LHC. Two freeze-out temperatures $T_f = 0.16$ GeV and $T_f = 0.14$ GeV are employed for demonstration. The former early freeze-out scenario is motivated by the possible break-down of hydrodynamic applicability due to the bulk viscous effects [51]. The net baryon distributions are calculated by taking into account the contributions of the hadron resonance [52] up to 2.5 GeV at freeze-out.

5.3.1 Net Baryon Distributions at RHIC

The net baryon rapidity distributions of the Au–Au collisions at $\sqrt{s_{NN}} = 200$ GeV with and without non-equilibrium corrections are shown in Fig. 5.2 for the freeze-out temperatures $T_f = 0.16$ GeV and $T_f = 0.14$ GeV. The parameters in the transport coefficients are set as $c_V = 1$, $c_{VW} = 0$ and $\tilde{c}_{AB} = 0$ for the moment. One can see that the hydrodynamic flow tends to carry the net baryon density to forward rapidity, broadening the flat region with relatively small baryon density near mid-rapidity in both equilibrium and non-equilibrium cases. The viscous hydrodynamic results are

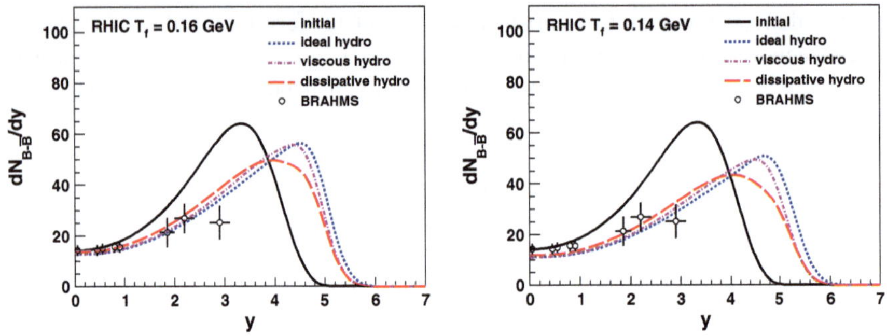

Fig. 5.2 The initial net baryon distributions based on the color glass theory (*solid line*) and the ones with the ideal (*dotted line*), the viscous (*dash-dotted line*) and the dissipative (*dashed line*) hydrodynamic effects at *(left)* $T_f = 0.16\,\text{GeV}$ and *(right)* $T_f = 0.14\,\text{GeV}$ for the Au–Au collisions at $\sqrt{s_{NN}} = 200\,\text{GeV}$. The experimental data points are the scaled results of the net proton rapidity distribution from the BRAHMS collaboration [7]

less flattened compared with the ideal hydrodynamic ones because the shear and the bulk viscosities prevent the expansion by effectively reducing the longitudinal pressure. The dissipative hydrodynamic results, which include the effects of baryon dissipation in addition to those of viscosities, differ from the viscous hydrodynamic results. The chemical potential gradient induces the baryon dissipation current into mid-rapidity region, further steepening the mid-rapidity valley. The data points are scaled results from the net proton distribution of the BRAHM experiments [7], which show relatively good agreement with the hydrodynamic results. It should be noted that quasi-quantitative discussion would be allowed here because the initial distribution is sensitive to the parameters in the color glass models and also the transverse dynamics is not taken into account.

The transparency of a collision can be quantified by baryon stopping. The rapidity loss is defined as $\langle \delta y \rangle = y_p - \langle y \rangle$ where y_p is the rapidity of incoming projectile and

$$\langle y \rangle = \int_0^{y_p} y \frac{dN_{B-\bar{B}}(y)}{dy} dy \bigg/ \int_0^{y_p} \frac{dN_{B-\bar{B}}(y)}{dy} dy, \tag{5.15}$$

is the mean rapidity of net baryon number. The ideal, the viscous and the dissipative hydrodynamic evolutions lead to the reduction in the average rapidity loss to $\langle \delta y \rangle$ = 2.09, 2.16 and 2.26 for $T_f = 0.16\,\text{GeV}$ and $\langle \delta y \rangle$ = 1.99, 2.06 and 2.19 for $T_f = 0.14\,\text{GeV}$, respectively, when that of the initial net baryon distribution is $\langle \delta y \rangle = 2.67$. The fact that hydrodynamic evolutions visibly reduce the rapidity loss suggests that the baryon stopping in RHIC would deviate *less* significantly from the extrapolation of AGS and SPS results at the formation of the hot medium, but the net baryon number is carried to the forward rapidity by the hydrodynamic medium interaction, effectively enhancing the observed transparency. It also indicates that the kinetic energy loss for the QGP production is larger than traditional expectations without the hydrodynamic effects at the time of medium formation, and part of the energy is transferred back to the net baryon component from the produced medium afterward.

The fact that the effects of the baryon dissipation current could be visible on an observable is important because it suggests that one would have to take the diffusion process into account to quantitatively understand the experimental data. It would also be useful in constraining the initial condition of the net baryon distribution which is not fully understood yet. The actual effect of the baryon diffusion could be larger because the current baryon charge conductivity is small as shown in Fig. 5.1. It is note-worthy that though shear and bulk viscosity and baryon dissipation seem to have similar effects on the net baryon rapidity distribution, the former enhances the baryon and the anti-baryon distributions individually while the latter only increases their difference, which implies that the contribution of baryon dissipation to the averaged distributions would be small.

Comparing the two freeze-out temperatures, the broadening effect is larger for the late freeze-out case because of the longer hydrodynamic evolution. This indicates that the late freeze-out tends to allow larger dissipative and viscous coefficients. It should be noted that the total baryon number is slightly smaller than the initial nuclei for $T_f = 0.14$ TeV because of the Cooper–Frye formulation of the freeze-out. The equation of state and the baryon fluctuation of kinetic theory and those of lattice QCD have to be identical at freeze-out to perfectly conserve the energy and the net baryon number because they should be reproduced in relativistic kinetic theory from the flow, the temperature, and the chemical potential. This would make it slightly more difficult to distinguish the freeze-out temperature dependence from the slopes of the net baryon distribution, but of course the freeze-out temperature dependence of the average rapidity losses remains unaffected.

5.3.2 Effects of Cross Terms

The effects of the cross terms in the equations of motion for the baryon dissipation current (5.6) are investigated next. Figure 5.3 shows the Soret effect in the QGP medium, which is induced by the linear thermo-diffusion term, at RHIC for $T_f = 0.16\,\text{GeV}$. $c_{VW} = 5, 0$ and -5 are employed for the cross coefficient. The semi-positive definite condition of the transport coefficient matrix is checked to be satisfied throughout the time evolution. One can see that the positive and the negative cross coefficients qualitatively lead to reduction and enhancement of the baryon diffusion effect, respectively. The results are consistent with the linear order analyses in Eq. (G.3) that the positive thermo-diffusion coefficient effectively reduces the charge conductivity. The magnitude of the thermo-diffusion effect, on the other hand, is relatively small in the present calculations because the cross coefficient becomes non-vanishing only in baryon rich region as a consequence of the matter-antimatter symmetry. This suggests that Soret effect might change the temperature and the chemical potential dependences of the transport properties, but it would be effective only at forward rapidity in high-energy heavy ion collisions.

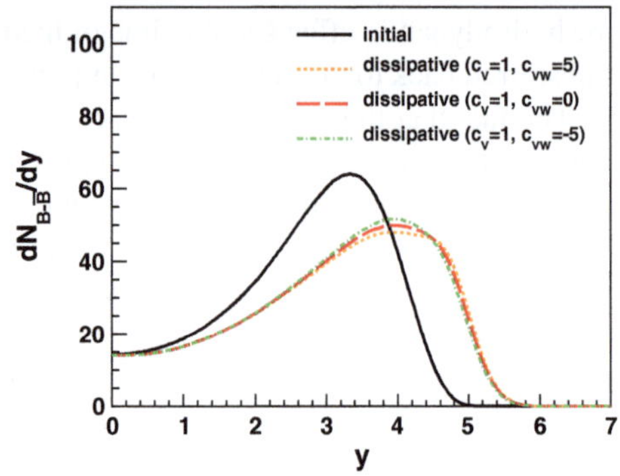

Fig. 5.3 The net baryon distribution at the initial stage based on the color glass theory (*solid line*) and the ones after dissipative hydrodynamic evolutions with the cross coefficients $c_{VW} = 5$ (*dotted line*), $c_{VW} = 0$ (*dashed line*) and $c_{VW} = -5$ (*dash-dotted line*) at $T_f = 0.16\,\text{GeV}$ for the Au–Au collisions at $\sqrt{s_{NN}} = 200\,\text{GeV}$

The second order cross terms could also affect the results because of the hierarchy in the magnitude of the dissipative currents mentioned in Sect. 5.1. Numerical estimations with finite $\tilde{c}_{AB}$ indicate that the bulk-shear cross term in the bulk pressure, and the baryon-shear and the baryon-bulk terms in the baryon dissipation can be relevant, assuming that the magnitude of the transport coefficients is roughly of the same order, i.e., $\tilde{c}_{AB} \sim \mathscr{O}(1)$. Note that in general there is much ambiguity in magnitude of the second order transport coefficients. The result is consistent with the fact that the shear pressure is larger than the bulk pressure, which in turn is larger than the baryon dissipation in relativistic heavy ion collisions.

5.3.3 Net Baryon Distributions at LHC

The prospects for the Pb–Pb collisions at $\sqrt{s_{NN}} = 2.76$ TeV in LHC experiment are shown in Fig5.4 for the two freeze-out temperatures. The net baryon distributions are still visibly modified by the hydrodynamic flow, effectively increasing the transparency in the collision. On the other hand, the effects of viscosities and dissipation are much smaller than those at RHIC. The final average rapidity losses after the ideal, the viscous and the dissipative hydrodynamic evolution are $\langle \delta y \rangle = 2.82$, 2.86 and 2.92 for $T_f = 0.16\,\text{GeV}$ and $\langle \delta y \rangle = 2.77$, 2.81 and 2.88 for $T_f = 0.14\,\text{GeV}$, respectively. The initial rapidity loss in the current parameter settings is $\langle \delta y \rangle = 3.36$. Comparing the early and the late freeze-out scenarios, the latter tends to have larger hydrodynamic effects as is found in the RHIC case.

The smaller dissipative effect could be understood as a result of the smaller spatial gradients in the chemical potential over temperature μ_B/T, the thermodynamic force to the baryon dissipation, at LHC. The shear and the bulk viscous effects are reduced for a different reason, because they do not directly respond to the gradient. The hydrostatic pressure P_0 increases with the collision energy more than the shear and the bulk pressures do because P_0 increases as $\sim T^4$ while the viscous coefficient $\sim T^3$. Thus the effects of viscous corrections in the effective pressure $P = P_0 + \Pi - \pi$

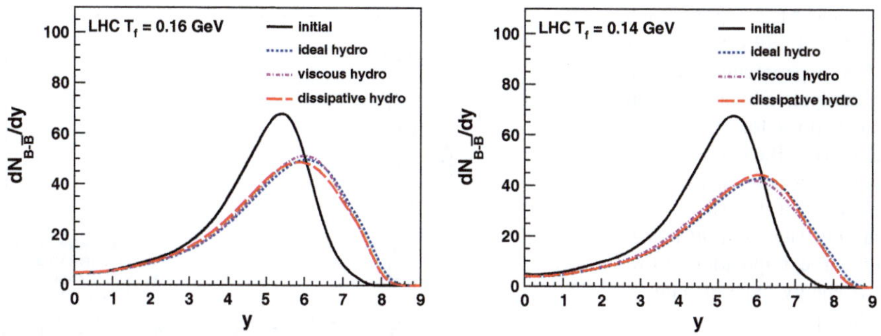

Fig. 5.4 The initial net baryon distributions based on the color glass theory (*solid line*) and the ones with the ideal (*dotted line*), the viscous (*dash-dotted line*) and the dissipative (*dashed line*) hydrodynamic effects at *(left)* $T_f = 0.16\,\text{GeV}$ and *(right)* $T_f = 0.14\,\text{GeV}$ for the Pb–Pb collisions at $\sqrt{s_{NN}} = 2.76\,\text{GeV}$

are decreased, consequently reducing the difference between the ideal and the viscous hydrodynamic results. It should be noted that the effect of baryon dissipation could be larger since the charge conductivity employed in the model is very small as mentioned earlier. Also the off-equilibrium corrections on the net baryon distribution might be underestimated due to the lack of explicit chemical potential dependence of the transport coefficients including the shear and the bulk viscosities in the current modeling.

Comparing the results with the ones at RHIC, the overall hydrodynamic effect is also found smaller at LHC. One of the reasons would be the fact that the mean rapidity loss in the initial net baryon distribution for the current parameter settings is large, i.e., the peak of the distribution is around $y \sim 5$ while the beam rapidity is $y \sim 8$. The pressure is larger at LHC, but the pressure gradient, which drives the net baryon current as convection, is relatively small in this rapidity region. Thus the actual hydrodynamic effect could be large enough to be measured in LHC experiments for the initial distributions with smaller mean rapidity loss or larger pressure gradient.

The rapidity losses at RHIC and LHC before and after the dissipation hydrodynamic evolution is plotted with those in AGS, SPS and RHIC experiments in Fig. 5.5. Bars for the RHIC data show experimental ambiguities. The dotted line denotes the extrapolation of the lower energy data. One can see that baryon stopping without hydrodynamic reduction is closer to the extrapolation.

5.4 Summary

A relativistic dissipative hydrodynamic model of the high-energy heavy ion collisions at finite density which takes account of shear viscosity, bulk viscosity and baryon dissipation with evolving flow is developed for the first time. The framework of

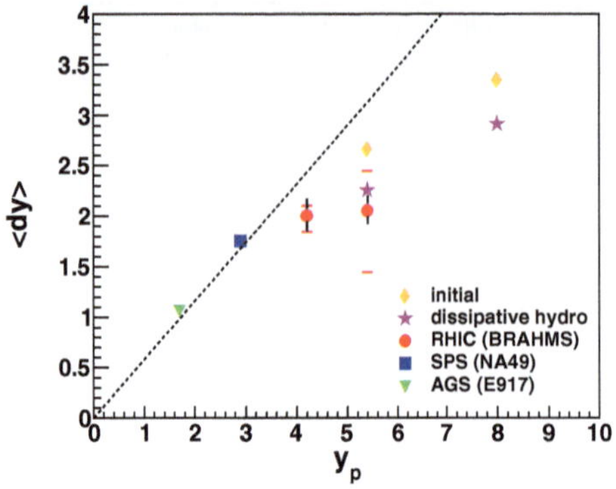

Fig. 5.5 Mean rapidity losses at AGS [4], SPS [5] and RHIC [7, 8] experiments and those in dissipative hydrodynamic calculations for RHIC and LHC. Bars show the ambiguity coming from the lack of forward rapidity data. The *dotted line* is a linear extrapolation of the low energy experimental data [7]

hydrodynamics is employed from the second order theory extended for systems with particle number changing processes. The initial conditions for the energy and the net baryon distributions are constructed from the color glass theory, and the EoS is employed from the Taylor expansion approach of the finite density lattice QCD to improve quantitative accuracy. It is found that the average rapidity loss for the baryon stopping is reduced during the hydrodynamic evolution, which would mean that the observed transparency of the collision at RHIC is effectively enhanced by the medium interaction. This suggests that more energy is available for the production of the hot medium at the initial stage than it was implied from the experimental data and that the strongly-coupled medium re-distribute part of the energy back to the net baryon components.

The net baryon distribution could also be sensitive to baryon dissipation as much as to viscosities. It should be noted that the current dissipative coefficients are rather small as shown in Fig. 5.1 and the actual diffusion effects can be larger. The effect of baryon dissipation could be important for explaining the experimental data and constraining the initial conditions for the net baryon distribution, which are not well known. One would need to introduce other observables such as the transverse momentum spectra of the net baryon number to constrain the transport coefficients of the hot QCD matter at finite baryon density from the collider experiments because of the ambiguities in the choice of initial conditions. The effects of the cross coefficients are also numerically investigated, and it is found that the thermo-diffusion effect, or the Soret effect, might modify the magnitude of the baryon diffusion, but the effect would be limited to the forward rapidity region because the baryon-heat cross conductivity vanishes for the baryon-free medium due to the matter-antimatter symmetry. The second-order cross terms among the dissipative currents of different magnitudes would also be important for the quantitative analyses. The late freeze-out is found to allow larger viscosities and baryon dissipation since the longer hydrodynamic evolution widens the mid-rapidity valley.

A possible source of overestimation for the hydrodynamic effects would be the lack of transverse expansion. The temperature and the chemical potential tend to be larger for the longitudinal geometry as the energy and the net baryon densities cannot spread into the transverse directions. It should be noted that while the transverse dependence of the net baryon distributions is experimentally implied to be small [30], the accelerated cooling would lead to the reduction in the effect of hydrodynamic evolution. This is partially taken into account by considering the early freeze-out scenario. The off-equilibrium corrections in the transverse expansion would be more non-trivial and worth investigating. Also the parametrization of the transport coefficients for finite density systems, especially the conductivities κ_V and κ_{VW}, needs to be improved through theoretical and experimental developments for more quantitative discussion.

References

1. W. Busza, A.S. Goldhaber, Nuclear stopping power. Phys. Lett. B **139**, 235 (1984)
2. F. Videbaek, O. Hansen, Baryon rapidity loss and mid-rapidity stacking in high-energy nucleus-nucleus collisions. Phys. Rev. C **52**, 2684 (1995)
3. L. Ahle et al. [E802 Collaboration], Proton and deuteron production in Au + Au reactions at 11.6/A GeV/c. Phys. Rev. C **60**, 064901 (1999)
4. B.B. Back et al. [E917 Collaboration], Baryon rapidity loss in relativistic Au+Au collisions. Phys. Rev. Lett. **86**, 1970 (2001)
5. H. Appelshauser et al. [NA49 Collaboration], Baryon stopping and charged particle distributions in central Pb + Pb collisions at 158 GeV per nucleon. Phys. Rev. Lett. **82**, 2471 (1999)
6. C. Blume [NA49 Collaboration], Centrality and energy dependence of proton, light fragment and hyperon production. J. Phys. G: Nucl. Part. Phys. **34**, S951 (2007)
7. I.G. Bearden et al. [BRAHMS Collaboration], Nuclear stopping in Au + Au collisions at $\sqrt{s_{NN}}$ = 200 GeV. Phys. Rev. Lett. **93**, 102301 (2004)
8. I.C. Arsene et al. [BRAHMS Collaboration], Nuclear stopping and rapidity loss in Au+Au collisions at $\sqrt{s_{NN}}$ = 62.4 GeV. Phys. Lett. B **677**, 267 (2009)
9. S.A. Bass, M. Bleicher, W. Cassing, A. Dumitru, H.J. Drescher, K. Eskola, M. Gyulassy, D. Kharzeev, Last call for RHIC predictions. Nucl. Phys. A **661**, 205 (1999)
10. G. Wolschin, Relativistic diffusion model. Eur. Phys. J. A **5**, 85 (1999)
11. G. Wolschin, Test of the Einstein relation in relativistic heavy ion collisions. Europhys. Lett. **47**, 30 (1999)
12. G. Wolschin, Diffusion and local deconfinement in relativistic systems. Phys. Rev. C **69**, 024906 (2004)
13. S. Soff, J. Randrup, H. Stoecker, N. Xu, Effects of strong color fields on baryon dynamics. Phys. Lett. B **551**, 115 (2003)
14. S.A. Bass, B. Mueller, D.K. Srivastava, Net baryon density in Au+Au collisions at the relativistic heavy ion collider. Phys. Rev. Lett. **91**, 052302 (2003)
15. K. Itakura, Y.V. Kovchegov, L. McLerran, D. Teaney, Baryon stopping and valence quark distribution at small x. Nucl. Phys. A **730**, 160 (2004)
16. V. Topor Pop, M. Gyulassy, J. Barrette, C. Gale, X.N. Wang, N. Xu, Baryon junction loops in HIJING/B anti-B v2.0 and the baryon/meson anomaly at RHIC. Phys. Rev. C **70**, 064906 (2004)
17. V. Topor Pop, M. Gyulassy, J. Barrette, C. Gale, Baryon anomaly and strong color fields in Pb+Pb collisions at 2.76A TeV at the CERN large Hadron collider. Phys. Rev. C **84**, 044909 (2011)
18. K.J. Eskola, H. Honkanen, H. Niemi, P.V. Ruuskanen, S.S. Rasanen, RHIC-tested predictions for low-p_T and high-p_T hadron spectra in nearly central Pb + Pb collisions at the CERN LHC. Phys. Rev. C **72**, 044904 (2005)

19. J.L. Albacete, Y.V. Kovchegov, Baryon stopping in proton-nucleus collisions. Nucl. Phys. A **781**, 122 (2007)
20. I.N. Mishustin, K.A. Lyakhov, Baryon deceleration by strong chromofields in ultrarelativistic nuclear collisions. Phys. Rev. C **76**, 011603 (2007)
21. K.A. Lyakhov, H.J. Lee, I.N. Mishustin, Baryon stopping and partonic plasma production by strong chromofields. Phys. Rev. C **84**, 055202 (2011)
22. Y. Mehtar-Tani, G. Wolschin, Baryon stopping as a new probe of geometric scaling. Phys. Rev. Lett. **102**, 182301 (2009)
23. Y. Mehtar-Tani, G. Wolschin, Baryon stopping and saturation physics in relativistic collisions. Phys. Rev. C **80**, 054905 (2009)
24. Y. Mehtar-Tani, G. Wolschin, Predictions for net-proton and net-kaon distributions at LHC energies. Phys. Lett. B **688**, 174 (2010)
25. Y.B. Ivanov, Baryon stopping in heavy-ion collisions at E_{lab} = 2–160 GeV/nucleon. Phys. Lett. B **690**, 358 (2010)
26. A. Monnai, Dissipative hydrodynamic effects on baryon stopping. Phys. Rev. C **86**, 014908 (2012)
27. L.D. McLerran, The color glass condensate and small x physics: four lectures. Lect. Notes Phys. **583**, 291 (2002)
28. J.-P. Blaizot, F. Gelis, Searching evidence for the color glass condensate at RHIC. Nucl. Phys. A **750**, 148 (2005)
29. F. Gelis, E. Iancu, J. Jalilian-Marian, R. Venugopalan, The color glass condensate. Ann. Rev. Nucl. Part. Sci. **60**, 463 (2010)
30. S.S. Adler et al. [PHENIX Collaboration], Identified charged particle spectra and yields in Au+Au collisions at $\sqrt{s_{NN}}$ = 200 GeV. Phys. Rev. C **69**, 034909 (2004)
31. B.B. Back et al. [PHOBOS Collaboration], Identified hadron transverse momentum spectra in Au+Au collisions at $\sqrt{s_{NN}}$ = 62.4 GeV. Phys. Rev. C **75**, 024910 (2007)
32. J.D. Bjorken, Highly relativistic nucleus-nucleus collisions: the central rapidity region. Phys. Rev. D **27**, 140 (1983)
33. L. Onsager, Reciprocal relations in irreversible processes. I. Phys. Rev. **37**, 405 (1931)
34. L. Onsager, Reciprocal relations in irreversible processes. II. Phys. Rev. **38**, 2265 (1931)
35. P. Colella, P.R. Woodward, The Piecewise Parabolic Method (PPM) for gas dynamical simulations. J. Comput. Phys. **54**, 174 (1984)
36. S. Borsanyi, G. Endrodi, Z. Fodor, A. Jakovac, S.D. Katz, S. Krieg, C. Ratti, K.K. Szabo, The QCD equation of state with dynamical quarks. JHEP **1011**, 077 (2010)
37. S. Borsanyi, Z. Fodor, S. D. Katz, S. Krieg, C. Ratti, K. Szabo, Fluctuations of conserved charges at finite temperature from lattice QCD. arXiv:1112.4416 [hep-lat] (2011)
38. P. Kovtun, D.T. Son, A.O. Starinets, Viscosity in strongly interacting quantum field theories from black hole physics. Phys. Rev. Lett. **94**, 111601 (2005)
39. M. Natsuume, T. Okamura, Causal hydrodynamics of gauge theory plasmas from AdS/CFT duality. Phys. Rev. D **77**, 066014 (2008) [Erratum-ibid. D **78**, 089902 (2008)]
40. A. Hosoya, M.A. Sakagami, M. Takao, Nonequilibrium thermodynamics In field theory: transport coefficients. Annals Phys. **154**, 229 (1984)
41. T. Hirano, Y. Nara, Hydrodynamic afterburner for the color glass condensate and the parton energy loss. Nucl. Phys. A **743**, 305 (2004)
42. T. Hirano, Y. Nara, Eccentricity fluctuation effects on elliptic flow in relativistic heavy ion collisions. Phys. Rev. C **79**, 064904 (2009)
43. A. Adil, H.-J. Drescher, A. Dumitru, A. Hayashigaki, Y. Nara, The Eccentricity in heavy-ion collisions from color glass condensate initial conditions. Phys. Rev. C **74**, 044905 (2006)
44. H.-J. Drescher, Y. Nara, Effects of fluctuations on the initial eccentricity from the color glass condensate in heavy ion collisions. Phys. Rev. C **75**, 034905 (2007)
45. H.-J. Drescher, Y. Nara, Eccentricity fluctuations from the color glass condensate at RHIC and LHC. Phys. Rev. C **76**, 041903 (2007)
46. D. Kharzeev, E. Levin, M. Nardi, QCD saturation and deuteron nucleus collisions. Nucl. Phys. A **730**, 448 (2004) [Erratum-ibid. A **743**, 329 (2004)]

47. A.D. Martin, R.G. Roberts, W.J. Stirling, R.S. Thorne, NNLO global parton analysis. Phys. Lett. B **531**, 216 (2002)
48. F. Cooper, G. Frye, Comment on the single particle distribution in the hydrodynamic and statistical thermodynamic models of multiparticle production. Phys. Rev. D **10**, 186 (1974)
49. E.V. Shuryak, M.A. Stephanov, When can long range charge fluctuations serve as a QGP signal? Phys. Rev. C **63**, 064903 (2001)
50. D. Teaney, J. Lauret, E.V. Shuryak, A hydrodynamic description of heavy ion collisions at the SPS and RHIC. arXiv:nucl-th/0110037 (2001)
51. G. Torrieri, B. Tomasik, I. Mishustin, Bulk viscosity driven clusterization of quark-gluon plasma and early freeze-out in relativistic heavy-ion collisions. Phys. Rev. C **77**, 034903 (2008)
52. K. Nakamura et al. [Particle Data Group Collaboration], Review of particle physics. J. Phys. G **37**, 075021 (2010)

Chapter 6
Summary and Conclusion

The collective dynamics of the quark-gluon plasma was studied in the framework of relativistic dissipative hydrodynamics in this thesis. The aim of the study is to describe the macroscopic space-time evolution of the QGP to explain the experimental data for understanding high-energy heavy ion collisions and obtaining insights into microscopic properties of the hot matter, e.g., the equation of state, the transport coefficients and the characteristic parameters of phenomenological models. For this purpose it is important to develop a relativistic dissipative hydrodynamic model analytically and numerically and to establish a clear approach to the non-equilibrium processes in the QGP. The beginning of the LHC experiments provides excellent opportunities to construct a more realistic theory, especially since the deviation from near equilibrium would be larger as QCD asymptotic freedom implies that the system would become a less strongly coupled. The summary of developments and prospects in the high-energy heavy ion physics is presented extensively from experimental and theoretical perspectives in Chap. 2.

In Chap. 3, we formulated relativistic dissipative hydrodynamics for systems with multiple conserved charges. One conventionally considers the second order corrections in terms of the dissipative currents for the entropy current. This introduces the time-like derivative of the dissipative current itself to the constitutive equations and promote them to dynamical equations, limiting the speed of propagation to finite and preserving causality. We showed that the Israel–Stewart theory, one of the most popular causal frameworks used for the analyses of heavy ion collisions, is applicable only for single component systems with elastic scatterings and is not suitable for analyzing a hot quark matter that involves particle creation and annihilation. It is found that one has to introduce the moment equations for charge conservation laws in addition to the conventional ones for energy-momentum conservation law for the completeness of the theory in the course of formulation. We derived full second-order dissipative hydrodynamic equations from the law of increasing entropy based on the momentum expansion method extended for systems with multiple conserved currents. The derived equations naturally have linear cross terms, which satisfy Onsager reciprocal relations. They can induce cross-coupling effects such as Soret effect, the chemical diffusion induced by the thermal gradients, or Dufour effect, the heat

A. Monnai, *Relativistic Dissipative Hydrodynamic Description of the Quark-Gluon Plasma*, Springer Theses, DOI: 10.1007/978-4-431-54798-3_6,

conduction caused by the chemical potential, which could be important in hot and dense systems. It was also shown that the formalism has all the second order terms allowed and required by kinetic theory and is mostly consistent with other hydrodynamic formalisms in the single-component and elastic scattering case, except for the vorticity–vorticity term in AdS/CFT formalism. The obtained equations are very generic and applicable not only to high-energy heavy ion collisions but also to early universe and ultra-cold atoms.

In Chap. 4, viscous hydrodynamic effects on the color glass condensate, a state of saturated gluons which gives description of the colliding nuclei, were estimated through numerical simulations. One of the latest results of the LHC experiment has shown that the charged particle multiplicity at mid-rapidity is about 30 % larger than that expected in pure CGC models, which had correctly estimated the quantity at RHIC. We estimated effects of viscous hydrodynamic evolution with the constitutive equations derived in Chap. 3 and found that the secondary interactions in a viscous hydrodynamic medium could be important in explaining the discrepancy between the CGC predictions and the LHC data. A new numerical algorithm was developed for the hydrodynamic analyses in the direction of the collision axis to estimate longitudinal space-time evolution from CGC-based initial conditions. The $(1 + 1)$-dimensional viscous hydrodynamic model would be a major step for $(3+1)$-dimensional analyses since the mixture of time-like and space-like derivatives is one of the difficult issues in numerical calculations for viscous fluids. It should be noted the hydrodynamic numerical code is a completely new one and not an extension of existing codes. The space-time evolutions of the QGP were estimated with a state-of-art equation of state from the first principle calculations of lattice QCD. The CGC initial conditions were employed from Nara adaptation of Monte–Carlo Kharzeev–Levin–Nardi model. The results show that the rapidity distributions are modified differently at RHIC and at LHC by the interplay of entropy production through non-equilibrium processes and entropy flux to forward rapidity region driven by non-boost invariant flow. For viscous coefficients close to the conjectured minimum values, the convection is effectively stronger than the viscous particle production and the distribution is flattened at RHIC while the effects of viscosity stood up with those of convection at LHC. This implies that if the hydrodynamic and the CGC parameters are tuned so the final distribution reproduce the experimental data at RHIC, additional particle production would take place at LHC. The results suggest that secondary interactions in the hydrodynamic stage could be essential for quantitative description of the multiplicity in the CGC models.

In Chap. 5, relativistic dissipative hydrodynamic effects on the QGP at finite baryon density were estimated. The net baryon number had been neglected in most hydrodynamic studies of high-energy heavy ion collisions, but it can be important in the context of precision physics especially at forward rapidity where the net baryon number of valence quarks in the shattered nuclei is conserved. Finite-density hydrodynamic models are also expected to become more important in beam energy scan experiments for the search of the QCD critical point. We developed a novel dissipative hydrodynamic model at finite baryon density and estimated baryon stopping, through which one estimates the energy available for medium production. It is found that the hydrodynamic effect visibly modifies the baryon stopping and that baryon diffusion process is non-negligible for the net baryon rapidity distribution. The relativistic dissipative hydrodynamic model developed in the study takes account of all the dissipative processes, i.e., shear viscosity, bulk viscosity and baryon dissipation. The initial condition was constructed from the valence quark distribution in the color glass theory and the finite-density equation of state was introduced from the lattice QCD calculations with Taylor expansion method. The results show that the hydrodynamic evolution effectively enhances the transparency of the collision. This implies that more energy is available for the production of the hot medium at the time of collision than what had been naïvely expected from experimental data because the hydrodynamic interaction gives part of the energy back to the net baryon components from the QGP medium in later stages. This would be a candidate explanation for the sudden increase in the collision transparency observed in the RHIC experiments compared with the linear extrapolation of the AGS and SPS results. The net baryon rapidity distribution is also found sensitive to baryon dissipation as much as to shear and bulk viscosities. Viscosity enhances the baryon stopping because the longitudinal pressure is effectively reduced with the expansion of the system. Diffusion also enhances the trend as it is induced by the chemical gradient into mid-rapidity due to the geometrical positions of the remnants of shattered nuclei. The fact that those effects are visible might suggest that the net rapidity distribution has valuable information on the transport coefficients at finite baryon density.

The numerical estimations were performed by integrating out the transverse dynamics. On the other hand, the lifetime of the QGP, for example, would be sensitive to the transverse expansion. Thus it would be important to develop a finite-density $(3 + 1)$-dimensional dissipative hydrodynamic model to better quantify the experimental data and to extract more information on the equation of state and the transport coefficients as well as on the initial conditions. This also motivates us to construct a grand model of high-energy heavy ion collisions where pre-collision physics, early thermalization, hydrodynamic evolution, freeze-out (Appendix A) and hadronic transport are integrated in a consistent manner. The on-going and coming collider experiments at RHIC and LHC would provide great opportunities and challenges for constructing realistic theoretical models. The relativistic dissipative hydrodynamic picture would provide an important foundation on which new theories for the quark-gluon plasma could be developed.

Appendix A
Viscous Corrections at Freeze-Out

The QCD fluid has to be converted into hadronic gas for the final stage of high-energy heavy ion collisions because (i) the applicability of hydrodynamics breaks down for lower energies and (ii) the flow is not an observable. The Cooper-Frye prescription (2.10), which is the most conventional framework for describing the freeze-out process, is subject to off-equilibrium corrections in two ways; one is modification on the profiles of the flow, the temperature and the chemical potentials, the other is distortion of the phase-space distribution itself. The former is implemented in dynamical evolutions of non-equilibrium fluids. The latter requires the estimation of the off-equilibrium distribution which can correctly preserve the off-equilibrium components of the energy-momentum tensor at the quark-hadron conversion and can create the right amount hadrons for the later stages.

In this appendix, viscous corrections at freeze-out via the modification of the phase-space distribution are investigated. Effects of both shear and bulk viscosities on p_T spectra and differential elliptic flow $v_2(p_T)$ are estimated with the distribution function improved from Ref. [1] based on the arguments in Chap. 3. Bulk viscosity in hydrodynamic evolution generally has a smaller effect compared with shear viscosity, but the effect could be large near the crossover region [2–6]. It is found that bulk viscous corrections can be as large as shear viscous ones at freeze-out due to subtleties in the consistent determination of distortion factors in the distribution function, even though bulk viscosity itself is small compared with shear viscosity and often neglected in numerical analyses.

A.1 Distortion of Phase-Space Distribution

The off-equilibrium corrections to the phase-space distribution (3.37) in the presence of net baryon charge is given as

$$\delta f^i = -f_0^i(1+\varepsilon f_0^i)(b_i p_i^\mu \varepsilon_\mu^B + p_i^\mu p_i^\nu \varepsilon_{\mu\nu}), \tag{A.1}$$

A. Monnai, *Relativistic Dissipative Hydrodynamic Description of the Quark-Gluon Plasma*, Springer Theses, DOI: 10.1007/978-4-431-54798-3,

where ε_μ^B and $\varepsilon_{\mu\nu}$ are coefficients of the expansion. $\varepsilon_{\mu\nu}$ is often considered to be traceless and the scalar term $\tilde{\varepsilon}$ is introduced instead [7]. The numbers of unknown variables are the same in both cases. If one considers a single component system such as a pion gas, the trace part of the tensor term is equivalent to a scalar term $\tilde{\varepsilon}$ since the trace part can be separated as $\mathrm{Tr}(\varepsilon_{\mu\nu})g_{\mu\nu}/4$ and the metric produces a momentum-independent scalar $p^\mu p^\nu g_{\mu\nu} = m^2$. On the other hand, this is not the case for multi-component systems. The distortion tensors are supposed to be macroscopic quantities in the sense that they do not depend on any single particle species but on all of them. Then the scalar correction cannot be identified as the trace of the distortion tensor since it has mass dependence, i.e., it is particle-species dependent:

$$
\begin{aligned}
\delta f^i_{\mathrm{tensor}} &= -f_0^i(1+\varepsilon f_0^i)p_i^\mu p_i^\nu \varepsilon_{\mu\nu} \\
&= -f_0^i(1+\varepsilon f_0^i)p_i^\mu p_i^\nu \left[\frac{\mathrm{Tr}(\varepsilon_{\mu\nu})}{4} g_{\mu\nu} + \left(\varepsilon_{\mu\nu} - \frac{\mathrm{Tr}(\varepsilon_{\mu\nu})}{4} g_{\mu\nu}\right)\right] \\
&= -f_0^i(1+\varepsilon f_0^i)\left[\frac{\mathrm{Tr}(\varepsilon_{\mu\nu})m_i^2}{4} + p_i^\mu p_i^\nu \tilde{\varepsilon}_{\mu\nu}\right],
\end{aligned} \tag{A.2}
$$

where $\tilde{\varepsilon}_{\mu\nu}$ is traceless. As discussed in Chap. 3, the finite-trace prescription is clearly preferred in multi-component/conserved current systems. The choice is also supported by the 16-component hadron resonance gas in numerical estimation as discussed later.

The Landau matching conditions $u_\mu \delta T^{\mu\nu} u_\nu = 0$ and $u_\mu \delta N_B^\mu = 0$, which ensure the thermodynamic stability in the first order theory, are introduced. The explicit expressions of ε_μ^B and $\varepsilon_{\mu\nu}$ are given as

$$\varepsilon_\mu^B = D_\Pi \Pi u_\mu + D_W W_\mu + D_V V_\mu, \tag{A.3}$$

$$\varepsilon_{\mu\nu} = (B_\Pi \Delta_{\mu\nu} + \tilde{B}_\Pi u_\mu u_\nu)\Pi + 2B_W u_{(\mu} W_{\nu)} + 2B_V u_{(\mu} V_{\nu)} + B_\pi \pi_{\mu\nu}, \tag{A.4}$$

where the distortion factors are

$$D_\Pi = 3(J_{40}J_{31}^B - J_{41}J_{30}^B)\mathscr{J}_3^{-1}, \tag{A.5}$$

$$B_\Pi = (J_{30}^B J_{30}^B - J_{40}J_{20}^{BB})\mathscr{J}_3^{-1}, \tag{A.6}$$

$$\tilde{B}_\Pi = 3(J_{41}J_{20}^{BB} - J_{30}^B J_{31}^B)\mathscr{J}_3^{-1}, \tag{A.7}$$

$$D_W = -2J_{31}^B \mathscr{J}_2^{-1}, \tag{A.8}$$

$$D_V = 2J_{41}\mathscr{J}_2^{-1}, \tag{A.9}$$

$$B_W = J_{21}^B \mathscr{J}_2^{-1}, \tag{A.10}$$

$$B_V = -J_{31}\mathcal{J}_2^{-1}, \tag{A.11}$$

$$B_\pi = \mathcal{J}_1^{-1}, \tag{A.12}$$

where

$$\begin{aligned}\mathcal{J}_3 = & 5J_{42}J_{30}^B J_{30}^B + 3J_{31}^B J_{40} J_{31}^B + 3J_{41}J_{41}J_{20}^{BB} \\ & - 3J_{31}^B J_{41} J_{30}^B - 3J_{41}J_{30}^B J_{31}^B - 5J_{42}J_{40}J_{20}^{BB},\end{aligned} \tag{A.13}$$

$$\mathcal{J}_2 = 2J_{31}^B J_{31}^B - 2J_{41}J_{21}^{BB}, \tag{A.14}$$

$$\mathcal{J}_1 = -2J_{42}. \tag{A.15}$$

The index B denotes a presence of the weight factor b_i in the definition of the moment (C.11). Here D_Π and D_V vanish upon taking the vanishing baryon chemical potential limit. The distortion factor for the shear stress tensor (A.12) has the same form as shown in Ref. [7], whereas those for the bulk pressure (A.5)–(A.7) are different from the ones in the reference because the distortion factors have explicit chemical potential dependence and the finite-trace tensor term is considered instead of the traceless tensor and the scalar terms.

A.1.1 Discussion

If the system has only one conserved current, the number of dissipative currents match that of unknowns within the framework of the conventional Grad's 14-moment method even if inelastic collisions are present. Then one may alternatively write the distortion of the distribution as

$$\delta y^i = p_i^\mu \varepsilon_\mu + p_i^\mu p_i^\nu \varepsilon_{\mu\nu}. \tag{A.16}$$

In this case, the Landau matching condition of the charge current vanishes upon taking the limit with J_{mn}^B. This could induce an ambiguity because the number of conditions for the determination of the distortion factors decreases. But if J_{mn}^B is expanded around the vanishing baryon chemical potential $\mu_B = 0$,

$$\tilde{J}_{mn} = 0 + \mu_B \left.\frac{\partial J_{mn}^B}{\partial \mu_B}\right|_{\mu_B=0} + \mathcal{O}(\mu_B^2), \tag{A.17}$$

is obtained. Then $u_\mu \delta N_B^\mu = 0$ is reduced to

$$\left[\frac{\partial J_{20}^B}{\partial \mu_B}\varepsilon_\| + \frac{\partial (J_{30}^B - J_{31}^B)}{\partial \mu_B}\varepsilon_{\|\|} + \frac{\partial J_{31}^B}{\partial \mu_B}\mathrm{Tr}(\varepsilon)\right]_{\mu_B=0} = 0, \tag{A.18}$$

which remains finite in the vanishing baryon density limit. This enables one to determine the distortion factors for the bulk pressure uniquely even after the limit is taken [1]. This expansion, however, may affect the derivation of the constitutive equations discussed in Chap. 3 because semi-positive definiteness of the transport coefficient matrices might not be satisfied a priori. Further investigation should be made to clarify whether the alternative expansion should be able to describe the single conserved current systems.

It should be noted that if one makes the minimum quadratic ansatz $\varepsilon_{\mu\nu} = C_1 \pi_{\mu\nu} + C_2 \Delta_{\mu\nu} \Pi$ [8], then the Landau condition would be violated because

$$u_\mu \delta T^{\mu\nu} u_\nu = -J_{41} \mathrm{Tr}(\varepsilon) \neq 0, \tag{A.19}$$

which implies that the full momentum expansion is preferred. Note that $\varepsilon_{\|\|}$'s in the matching conditions vanish in the quadratic ansatz since $\varepsilon_{\mu\nu}$ is defined perpendicular to u^μ. The magnitude of bulk viscous correction is naïvely expected to be smaller than that of shear viscosity, because $C_2 = -\frac{2}{5} C_1$ holds.

It is also important to consider multi-component systems explicitly for estimations of the viscous corrections (A.5)–(A.12) because the deviation of the distribution for the ith particle species δf_i should generally depend on whether it is the only component or one of the components in a multi-component system. This comes from the fact that J_{mn}'s, and thus the distortion factors, may include information of all particle species, making them thermodynamic variables. It is interesting note that the difference becomes negligible in a special case where one considers the first order theory and assumes that shear viscosity η is proportional to the entropy density s, because in the Boltzmann approximation one has $s = J_{42}/T^3$ and the moment factor cancels out with $B_2 = 1/J_{42}$, i.e., $\delta f^i_{\mathrm{shear}} = B_2 \times 2\eta \nabla_{\langle\mu} u_{\nu\rangle} \times p_i^\mu p_i^\nu \propto T^{-3} \nabla_{\langle\mu} u_{\nu\rangle} \times p_i^\mu p_i^\nu$.

A.2 The Model

Numerical estimation of the distortion factors in δf is considered. The 16-component hadron resonance gas is employed as a model equation of state to keep consistency with kinetic theory. The model takes account of mesons and baryons up to $\Delta(1232)$ in mass. Hereafter the Landau frame is chosen and the system is assumed to be baryon free. The models for shear viscosity η and bulk viscosity ζ are employed from Refs. [9] and [10–12] as $\eta = s/4\pi$ and $\zeta = \alpha(\frac{1}{3} - c_s^2)^2 \eta$, respectively. The factor α in bulk viscosity is set to 15 unless there is further notification. Figure A.1 shows the temperature dependence of η/s and ζ/s in the hadronic resonance gas model. In the temperature range relevant to relativistic heavy ion collisions around $0.1 < T < 0.2$ GeV, ζ/s is much smaller than η/s. This does not necessarily mean that the bulk viscous effects are small since the distortion factors in front of the dissipative currents could be large.

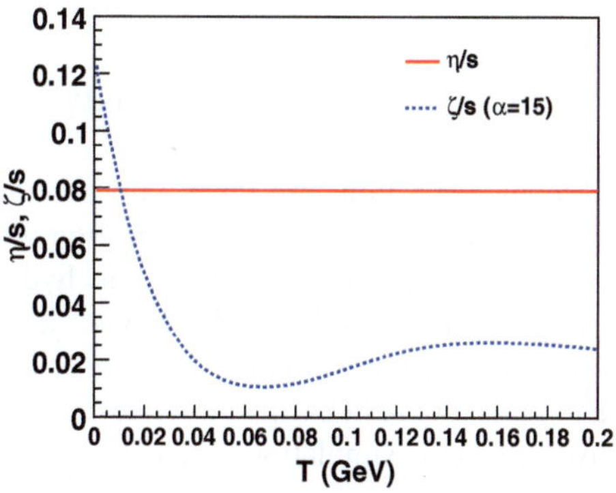

Fig. A.1 η/s (*solid*) and ζ/s with $\alpha = 15$ (*dotted*) as functions of the temperature in the hadron resonance gas model

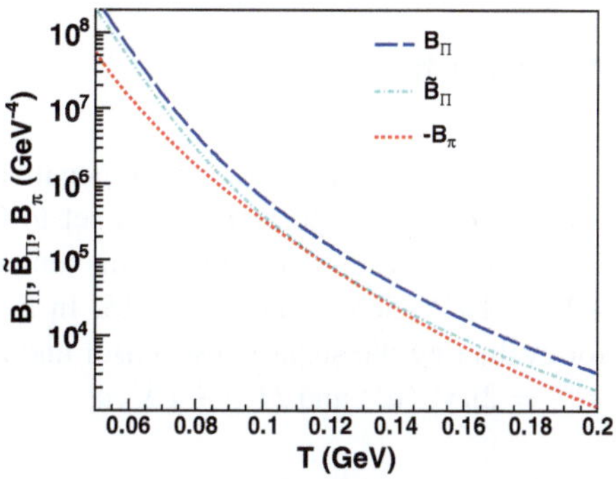

Fig. A.2 The distortion factors in $\varepsilon_{\mu\nu}$ for the bulk pressure, B_Π (*dashed*), and $\tilde{B}_\Pi$ (*dash-dotted*), and that for the shear stress tensor B_π (*solid*) as functions of the temperature

Figure A.2 shows the numerical results of the distortion factors B_Π, $\tilde{B}_\Pi$, and B_π in δf as functions of the temperature in the hadronic resonance gas model. It is interesting to note that the numerical analyses suggest that one should consider a finite trace tensor correction term instead of a scalar term in the 16-component hadron resonance gas model, since the combination of the scalar term and the traceless tensor term leads to divergence in the distortion factors due to a sign change in the denominator (A.13) at a temperature below 0.2 GeV.

The Cooper-Frye formula [13], as mentioned in previous chapters, converts the macroscopic variables into microscopic distributions, allowing one to compare hydrodynamic results with experimental data. p_T spectra of the ith particle is given as

$$\frac{dN_i}{d^2 p_T dy} = \frac{g_i}{(2\pi)^3} \int_\Sigma p_i^\mu d\sigma_\mu (f_0^i + \delta f^i), \tag{A.20}$$

by explicitly separating the local equilibrium distribution f_0 and the off-equilibrium correction δf. The elliptic flow coefficient is the coefficient of the second harmonics $\cos(2\phi_p)$ in a Fourier expansion of the azimuthal momentum distribution. The differential elliptic flow is expressed as

$$v_2(p_T) = \frac{\int dy \int_0^{2\pi} d\phi_p \cos(2\phi_p) \frac{dN_i}{d\phi_p p_T dp_T dy}}{\int dy \int_0^{2\pi} d\phi_p \frac{dN_i}{d\phi_p p_T dp_T dy}}. \tag{A.21}$$

Viscous corrections are taken into account via the distortion in the flow and the temperature fields and the modification of the distribution function. Here the focus is put on the latter since the viscous hydrodynamic radial flow would be required for the former. Profiles of the flow u^μ and the freeze-out hypersurface $d\sigma_\mu$, which are necessary for estimating the Cooper-Frye formula (A.20), are taken from (a) the Bjorken model [14], (b) a blast wave model [15], and (c) a (3+1)-dimensional ideal hydrodynamic simulation [16, 17].

A.3 Results

p_T spectra and elliptic flow of the negative pion ($m = 0.139$ GeV) are investigated. The freeze-out temperature T_f is set to 0.16 GeV, which is sufficiently near the QCD crossover temperature. At this temperature $\eta = 1.31 \times 10^{-3}$ GeV^3 and $\zeta = 4.37 \times 10^{-4}$ GeV^3 when $\alpha = 15$. In this study the Navier-Stokes limit is taken for evaluating the shear stress tensor and the bulk pressure from the flow field, i.e., $\pi^{\mu\nu} = 2\eta\nabla^{\langle\mu}u^{\nu\rangle}$ and $\Pi = -\zeta\nabla_\mu u^\mu$.

A.3.1 Bjorken Flow

The boost-invariant flow is considered as the first example. In the relativistic and cyllindrical frame, the Bjorken flow is written as

$$u^\tau = 1,\ u^r = u^\phi = u^{\eta_s} = 0. \tag{A.22}$$

The radius of the nuclei and the freeze-out time for the Cooper-Frye formula are set as $R_0 = 10.0$ fm and $\tau = 7.5$ fm, respectively. Elements of freeze-out hypersurface only have the radial component:

$$d\sigma_\tau = \tau d\eta_s r dr d\phi,\ d\sigma_r = d\sigma_\phi = d\sigma_{\eta_s} = 0. \tag{A.23}$$

It is note-worthy that the Cooper-Frye formula for this model can be analytically expressed in Boltzmann approximation. Details can be found in Sect. A.4.

Figure A.3 shows the particle spectra with corrections from bulk viscosity and shear viscosity. The mean transverse momentum, $\langle p_T\rangle$, is decreased by bulk viscosity and increased by shear viscosity for the Bjorken flow. This can be interpreted from the arguments on the sign of the shear stress tensor and the bulk pressure as follows. The bulk pressure works as a negative pressure, i.e., the pressure in the energy

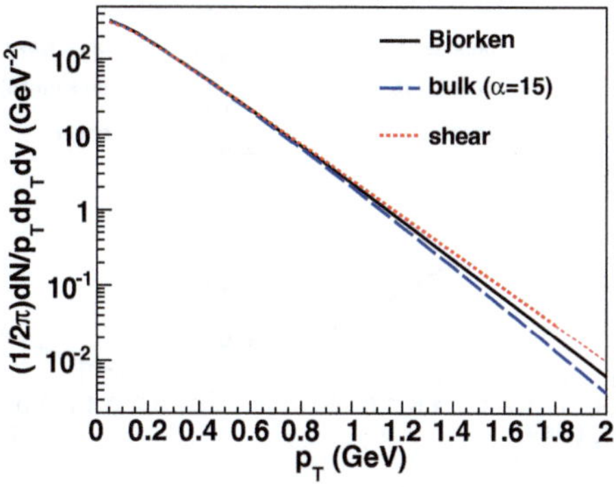

Fig. A.3 Viscous corrections to p_T spectra with Bjorken flow. *Solid*, *thick dashed*, and *thick dotted lines* are, respectively, the results without any viscous corrections, with the effect of bulk viscosity ($\alpha = 15$), and with the effect of shear viscosity. *Thin lines* indicate that the absolute value of the ratio of the correction to the ideal spectrum becomes greater than 0.5

momentum tensor is effectively reduced to $P - |\Pi|$ because the system is expanding in the longitudinal direction and the thermodynamic force $\partial_\mu u^\mu = 1/\tau$ is positive. As a result, the number of particles with lower momenta increases. Likewise, shear viscosity is naïvely expected to enhance $\langle p_T \rangle$ in the mid-rapidity region because the pressure in the radial direction is increased by the decrease of the longitudinal pressure due to the fact that the shear stress tensor is traceless.

A.3.2 Flow from Blast Wave Model

As a blast wave model, the parametrization of Ref. [15] is employed:

$$u^r = u_0 \frac{r}{R_0}[1 + u_2 \cos(2\phi)]\Theta(R_0 - r), \tag{A.24}$$

$$u^\tau = \sqrt{1 + (u^r)^2}, \tag{A.25}$$

$$u^\phi = u^{\eta_s} = 0, \tag{A.26}$$

where $u_0 = 0.55$ and $u_2 = 0.2$. The radius of the nuclei and the freeze-out time are set to $R_0 = 7.5$ fm and $\tau = 5.25$ fm. The profile of the freeze-out hypersurface is the same as Eq. (A.23) in the Bjorken model.

Again bulk viscosity lowers the spectrum as shown in Fig. A.4. The elliptic flow parameter $v_2(p_T)$, on the other hand, is *enhanced* by bulk viscosity. The counter-intuitive result could be interpreted as that the slope of differential $v_2(p_T)$ could be given as $dv_2(p_T)/dp_T \simeq v_2/\langle p_T \rangle$ [18]. Note that the average v_2 is not much affected by the viscous correction in the case of the blast wave model. Effects of shear viscosity can be explained likewise.

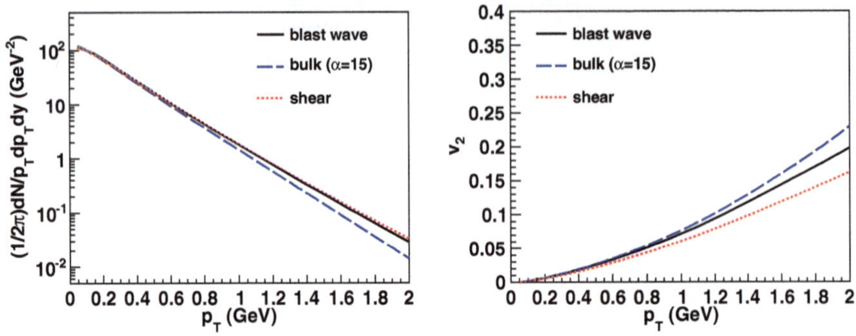

Fig. A.4 Viscous corrections to (*left*) p_T spectra and (*right*) $v_2(p_T)$ in the blast wave model. *Solid*, *thick dashed*, and *thick dotted lines* are, respectively, the results without any viscous corrections, with the effect of bulk viscosity ($\alpha = 15$), and with the effect of shear viscosity

A.3.3 (3+1)-Dimensional Ideal Hydrodynamic Flow

Lastly, the effect of shear and bulk viscosity is estimated with a publicly available (3+1)-dimensional ideal hydrodynamic flow profile from the Au–Au collisions at $\sqrt{s_{NN}} = 200$ GeV with the impact parameter $b = 7.2$ fm [16, 17]. The EoS in the hydrodynamic model for the hadron phase is the same as the one considered here, i.e., that of the hadronic resonance gas with particles up to $\Delta(1232)$ in mass. Validity of off-equilibrium corrections is confirmed by checking $|\Pi/P|$ and $|\pi^{\mu\nu}/T_0^{\mu\nu}|$ at each freeze-out position to be less than unity.

One can see in Fig. A.5 that *both* shear and bulk effects lower the spectrum in the mid-high p_T region. The enhancement of $v_2(p_T)$ by bulk viscosity is due to the decrease in mean p_T of the spectrum as discussed in Sect. A.3.2. The non-triviality lies in the correction of shear viscosity; it lowers the p_T-spectrum and still suppresses $v_2(p_T)$. This behavior was also observed in a (2+1)-dimensional viscous hydrodynamic model [19] in the context of flow modification. It could be due to the presence of stronger radial flow gradients. Note that the viscous corrections in these calculations may have been overestimated because the viscous correction to the flow is neglected. Viscosity tends to make the system smeared and consequently the thermodynamic forces, in particular, derivatives of the flow are expected to become smaller in the case of viscous hydrodynamic flow. This would reduce the amount of corrections considered here.

Figure A.6 shows comparison to p_T spectra and $v_2(p_T)$ which assumes the minimum quadratic ansatz mentioned in Sect. A.1.1. Neither the amount of the correction nor its p_T dependence seems to be similar to each other. The minimum quadratic ansatz tends to underestimates the effects of bulk viscosity. This indicates that the full off-equilibrium distortion factors in δf should be considered without making approximation as discussed in Sect. 5.1 to correctly calculate the distortion of particle spectra due to bulk viscosity.

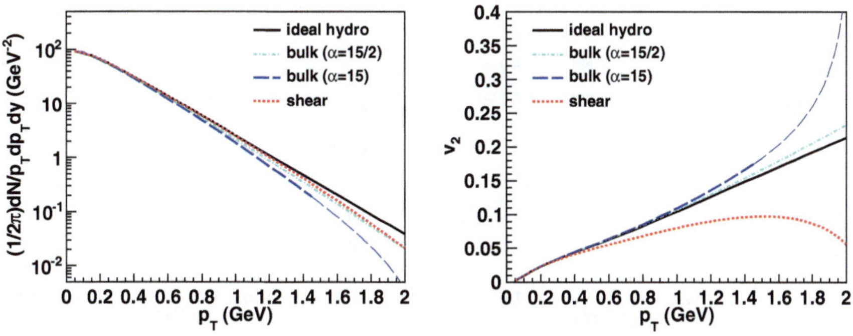

Fig. A.5 Viscous corrections to (*left*) p_T spectra and (*right*) $v_2(p_T)$ with ideal hydrodynamic flow. *Solid*, *thick dash-dotted*, *thick dashed*, and *thick dotted lines* are, respectively, the results without any viscous corrections, with the effect of bulk viscosity with $\alpha = 15/2$ and with $\alpha = 15$, and with the effect of shear viscosity. *Thin lines* indicate that the absolute value of the ratio of the correction to the ideal spectrum becomes larger than 0.5

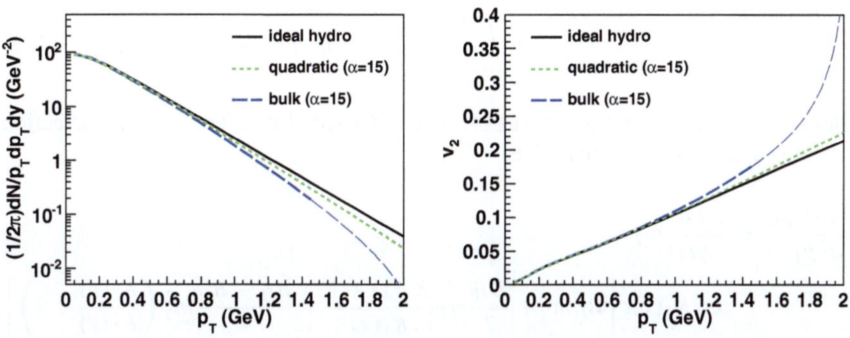

Fig. A.6 Viscous corrections to (*left*) p_T spectra and (*right*) $v_2(p_T)$ with (*thick dotted*) or without (*thick dashed*) the minimum quadratic approximation. *Solid line* is the result from an ideal hydrodynamic model. *Thin lines* indicate that the absolute value of the ratio of the correction to the ideal spectrum becomes larger than 0.5

Finally when *both* shear viscosity and bulk viscosity are considered, the slope of the particle spectra becomes steeper but that of $v_2(p_T)$ becomes flatter compared with that of the ideal distribution up to $p_T \sim 1$ GeV. The reason for the latter would be the accidental cancellation between the effect of bulk viscosity and that of shear viscosity in the low p_T region (Fig. A.7).

A.4 Analytic Expressions of Particle Spectra

The corrections of shear viscosity and bulk viscosity in the Cooper-Frye formula for the Bjorken model can be analytically expressed in the Boltzmann approximation following the discussion in Ref. [15]. Since the momentum p^τ is given by $m_T \cosh(y - \eta_s)$ where m_T is the transverse mass, the particle spectra is written as

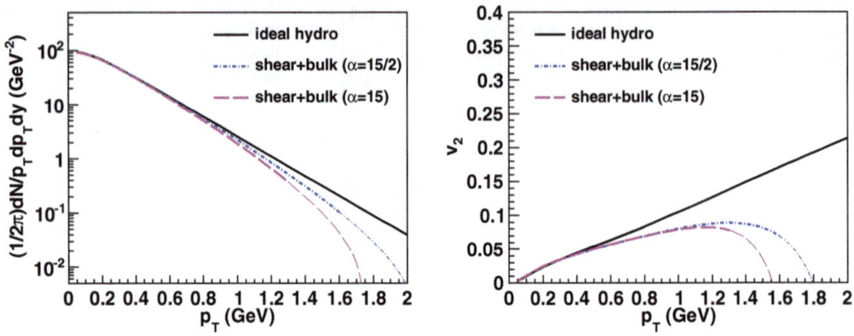

Fig. A.7 Viscous corrections of *both* shear viscosity and bulk viscosity to (*left*) p_T spectra and (*right*) $v_2(p_T)$. *Solid*, *thick dash-dotted*, and *thick dashed lines* are, respectively, the results without any viscous corrections, with the effect of both shear and bulk viscosities with $\alpha = 15/2$, and with $\alpha = 15$. *Thin lines* indicate that the absolute value of the ratio of the correction to the ideal spectrum becomes larger than 0.5

$$\frac{dN}{d^2 p_T dy} = \frac{g}{(2\pi)^3} \int_0^{R_0} r dr \int_0^{2\pi} d\phi \int_{-\infty}^{\infty} \tau d\eta_s m_T \cosh(y - \eta_s) f. \tag{A.27}$$

Here we dropped the index i of particle species for simplicity. The viscous corrections are now analytically expressed as

$$\begin{aligned} \frac{d\delta N_{\text{bulk}}}{d^2 p_T dy} &= \frac{g\pi R_0^2}{(2\pi)^3} m_T \\ &\times 2K_1(x)\zeta \left[B_\Pi m^2 + \frac{D_\Pi}{2} m_T \frac{K_2(x)+1}{K_1(x)} + \frac{\tilde{B}_\Pi - B_\Pi}{4} m_T^2 \left(\frac{K_3(x)}{K_1(x)} + 3 \right) \right], \end{aligned} \tag{A.28}$$

for bulk viscosity and

$$\frac{d\delta N_{\text{shear}}}{d^2 p_T dy} = \frac{g\pi R_0^2}{(2\pi)^3} m_T \times 2K_1(x) \frac{2\eta}{3} B_\pi \left[\frac{m_T^2}{2} \left(\frac{K_3(x)}{K_1(x)} - 1 \right) - p_T^2 \right], \tag{A.29}$$

for shear viscosity [15], where $x = m_T/T$ and the modified Bessel function is defined as

$$K_n(x) = \int_0^{\infty} e^{-x\cosh(t)} \cosh(nt) dt. \tag{A.30}$$

Here the bulk pressure and the shear stress tensor are estimated in first order theory.

A.5 Summary

The viscous corrections on the phase space distribution for a relativistic gas are estimated in multi-component systems. It is found that generalization to a multi-component gas is a subtle issue. Firstly, the trace part of the tensor term and the scalar term in the distortion of the distribution function are equivalent for single component systems, but not for multi-component systems. This point is also discussed in Chap. 3. Numerical calculations further support that the distortion tensor should have finite trace in the case of the 16-component hadron resonance gas. Secondly, the deviation of the distribution for the i-th gas component δf_i is generally different depending on whether it is the only component or one of the components. This comes from the fact that the distortion factors in ε_μ and $\varepsilon_{\mu\nu}$, like thermodynamic variables, include information of all the components in the system.

The Cooper-Frye formula is a necessary part of the hydrodynamic models of relativistic high-energy heavy ion collisions because the flow has to be converted into particle spectra at freeze-out. It enables one to compare the hydrodynamic results with experimental data, or to estimate further time evolution of the hadronic matter in a cascade model. Non-equilibrium effects appear in the formula as the variation of the flow and as the modification of the distribution. Here the focus is put only on the latter and numerical estimation of the viscous corrections on p_T spectra and elliptic flow $v_2(p_T)$ are performed. The profiles of the flow and the freeze-out hypersurface are taken from the Bjorken model with cylindrical geometry, a blast wave model with azimuthally anisotropic flow, and a (3+1)-dimensional ideal hydrodynamic model. It is found that corrections of bulk viscosity due to the distortion of the distribution have a visible effect on particle spectra and elliptic flow coefficient $v_2(p_T)$. The results imply the importance of bulk viscosity in arguing extraction of the transport coefficients from experimental data.

Quantitatively speaking, the viscous effects might have been overestimated because the derivatives of ideal hydrodynamic flow are generally larger than those of viscous hydrodynamic flow. Considering the fact that shear viscosity also has non-trivial effects on particle spectra depending on the flow profile, a full (3+1)-dimensional viscous hydrodynamic flow is necessary to see more realistic behavior of p_T spectra and $v_2(p_T)$ at freeze-out.

A.5 Summary

Appendix B
Thermodynamics for Relativistic Systems

The basics of relativistic thermodynamics are summarized in this appendix. The temperature, the chemical potentials and the equilibrium quantities—the energy density e_0, the hydrostatic pressure P_0, the charge densities n_{J0} ($J = 1, ..., N$) and the entropy density s—are closely related via thermodynamics and the equation of state. There are three laws of thermodynamics; the first law is

$$ds = \frac{1}{T} de_0 - \frac{\mu_J}{T} dn_{J0}. \tag{B.1}$$

This is another expression of energy-momentum conservation. The summation symbol for charges is abbreviated for simplicity. The second law of thermodynamics is

$$ds - \frac{dQ}{T} \geq 0, \tag{B.2}$$

where Q is the heat, or equivalently in much familiar form for relativistic hydrodynamics,

$$\partial_\mu s^\mu \geq 0, \tag{B.3}$$

which also is known as the law of increasing entropy. The third law of thermodynamics states that the entropy has to be vanishing at the absolute zero temperature,

$$s(T = 0) = 0. \tag{B.4}$$

At the vanishing chemical potential limit, the following relations are useful for calculating one thermodynamic quantity from another:

$$sT = e_0 + P_0, \tag{B.5}$$

A. Monnai, *Relativistic Dissipative Hydrodynamic Description of the Quark-Gluon Plasma*, Springer Theses, DOI: 10.1007/978-4-431-54798-3,

$$s = \int^{T} dT' P_0(T'), \tag{B.6}$$

$$\frac{P_0}{T^4} = \int^{T} dT' \frac{e_0 - 3P_0}{T^5}, \tag{B.7}$$

where the integrand on the last relation is the dimensionless trace anomaly, which is a direct observable in lattice QCD calculations.

The Gibbs-Duhem relation, which is derived as a Legendre transformation of the first law of thermodynamics, reads

$$dP_0 = sdT + n_{J0} d\mu_J \tag{B.8}$$

$$= -(e_0 + P_0) T d\frac{1}{T} + n_{J0} T d\frac{\mu_J}{T}, \tag{B.9}$$

which implies

$$\left(\frac{\partial P_0}{\partial e_0}\right)_{n_{J0}} de_0 = -(e_0 + P_0) T d\frac{1}{T}, \tag{B.10}$$

$$\left(\frac{\partial P_0}{\partial n_{J0}}\right)_{e_0} dn_{J0} = n_{J0} T d\frac{\mu_J}{T}. \tag{B.11}$$

It is quite useful to derive hydrodynamic relations from here using the conservation laws of hydrodynamic systems. Energy-momentum conservation reads

$$De_0 = -(e_0 + P_0)\nabla_\mu u^\mu - \Pi \nabla_\mu u^\mu + \pi^{\mu\nu} \nabla_{\langle\mu} u_{\nu\rangle}, \tag{B.12}$$

thus

$$\begin{aligned} D\frac{1}{T} &= \left(\frac{\partial P_0}{\partial e_0}\right)_{n_{B0}} \frac{1}{T} \left[\nabla_\mu u^\mu + \frac{1}{e_0 + P_0} (\Pi \nabla_\mu u^\mu + \pi^{\mu\nu} \nabla_{\langle\mu} u_{\nu\rangle})\right] \\ &= c_e^2 \frac{1}{T} \nabla_\mu u^\mu + \frac{c_e^2}{(e_0 + P_0)T} (\Pi \nabla_\mu u^\mu + \pi^{\mu\nu} \nabla_{\langle\mu} u_{\nu\rangle}), \end{aligned} \tag{B.13}$$

where $c_e^2 = (\partial P_0 / \partial e_0)_{n_{J0}}$ is the squared sound velocity if the system did not have conserved charges. Likewise, net charge conservation

$$Dn_{J0} = -n_{J0} \nabla_\mu u^\mu - \nabla_\mu V^\mu + V^\mu Du_\mu, \tag{B.14}$$

yields

$$D\frac{\mu_J}{T} = -\left(\frac{\partial P_0}{\partial n_{J0}}\right)_{e_0}\frac{1}{T}\left[\nabla_\mu u^\mu + \frac{1}{n_{J0}}(\nabla_\mu V^\mu - V^\mu D u_\mu)\right]$$
$$= -c_{n_J}^2\frac{1}{T}\nabla_\mu u^\mu - \frac{c_{n_J}^2}{n_{J0}T}(\nabla_\mu V^\mu - V^\mu D u_\mu), \qquad \text{(B.15)}$$

where $c_{n_J}^2 = (\partial P_0/\partial n_{J0})_{e_0}$ characterizes the finite density correction to the squared sound velocity. The space-like projection of energy momentum conservation also leads to

$$\nabla^\mu\frac{1}{T} + \frac{1}{T}Du^\mu = \frac{n_{J0}}{e_0 + P_0}\nabla^\mu\frac{\mu_J}{T}$$
$$+ \frac{1}{(e_0 + P_0)T}(\nabla^\mu \Pi - \Pi D u^\mu + \pi^{\mu\nu} D u_\nu + \Delta^{\mu\nu}\nabla^\rho \pi_{\nu\rho}), \qquad \text{(B.16)}$$

which suggests that at the first order in ideal hydrodynamic limit the directions of the energy dissipation and the charge dissipation are the same. This is natural as there are no leaks of conserved quantities from the flow in the limit.

It should be noted that the sound velocity is given as

$$c_s^2 = \left(\frac{\partial P_0}{\partial e_0}\right)_{s/n_{J0}} = \left(\frac{\partial P_0}{\partial e_0}\right)_{n_{J0}} + \frac{n_{J0}}{e_0 + P_0}\left(\frac{\partial P_0}{\partial n_{J0}}\right)_{e_0} = c_e^2 + \frac{n_{J0}}{e_0 + P_0}c_{n_J}^2. \qquad \text{(B.17)}$$

Appendix C
Relativistic Kinetic Theory

Relativistic kinetic theory gives macroscopic description of dilute systems from the microscopic phase-space distribution f. In the (3+1)-dimensional space-time, the energy-momentum tensor and the conserved net charge number current are expressed as the second and the first moments of distribution as

$$T^{\mu\nu} = \sum_i \int \frac{dp^3}{(2\pi)^3 E_i} p^\mu p^\nu f^i, \quad (C.1)$$

$$N_J^\mu = \sum_i \int \frac{q^i_J dp^3}{(2\pi)^3 E_i} p^\mu f^i, \quad (C.2)$$

where $p^\mu = (E, p^m)$ $(m = x, y, z)$ and the summation over particle species i is taken. q^i_J is the quantum number of particle species i for the Jth conserved charge. The degeneracy g_i is not written for the moment for simplicity. Here the energy density is

$$T^{00} = \sum_i \int \frac{dp^3}{(2\pi)^3} E_i f^i. \quad (C.3)$$

The momentum density and the energy flux are equivalent since the energy-momentum tensor is symmetric:

$$T^{m0} = T^{0m} = \sum_i \int \frac{dp^3}{(2\pi)^3} p_i^m f^i = \sum_i \int \frac{dp^3}{(2\pi)^3} E_i v^m f^i, \quad (C.4)$$

where $v_i^m = p_i^m / E_i$ is the velocity. The momentum flux, or the stress tensor, is expressed as

A. Monnai, *Relativistic Dissipative Hydrodynamic Description of the Quark-Gluon Plasma*, Springer Theses, DOI: 10.1007/978-4-431-54798-3,

$$T^{mn} = \sum_i \int \frac{dp^3}{(2\pi)^3} p_i^m v_i^n f^i. \tag{C.5}$$

Likewise, the number density and the number flux are

$$N_J^0 = \sum_i \int \frac{q_J^i dp^3}{(2\pi)^3} f^i, \tag{C.6}$$

$$N_J^m = \sum_i \int \frac{q_J^i dp^3}{(2\pi)^3} v_i^m f^i. \tag{C.7}$$

The entropy current is given in

$$s^\mu = -\sum_i \int \frac{dp^3}{(2\pi)^3 E_i} p^\mu [f_i \ln f_i - \varepsilon^{-1}(1+\varepsilon f_i)\ln(1+\varepsilon f_i)]. \tag{C.8}$$

The thermodynamic relation $e_0 + P_0 - \mu_J n_{J0} = sT$ is recovered in the equilibrium limit.

The nth moments of the phase space distribution and their weighted counterparts are defined as

$$I^{\mu_1\mu_2...\mu_n} = \sum_i \int \frac{dp^3}{(2\pi)^3 E_i} p^{\mu_1} p^{\mu_2} ... p^{\mu_n} f^i, \tag{C.9}$$

$$I^{\mu_1\mu_2...\mu_n}_{j_1, j_2 ..., j_N} = \sum_i \int \frac{(q_{J_1} q_{J_2} ... q_{J_N}) dp^3}{(2\pi)^3 E_i} p^{\mu_1} p^{\mu_2} ... p^{\mu_n} f^i. \tag{C.10}$$

It is convenient to define additional moments of the equilibrium distribution when momentum expansion of the distribution functions is considered with quantum statistics ε:

$$J^{\mu_1...\mu_n}_{j_1,...,j_N} = \sum_i \int \frac{(q_{J_1} ... q_{J_N}) dp^3}{(2\pi)^3 E_i} p^{\mu_1} ... p^{\mu_n} f_0^i (1+\varepsilon f_0^i), \tag{C.11}$$

$$K^{\mu_1...\mu_n}_{j_1,...,j_N} = \sum_i \int \frac{(q_{J_1} ... q_{J_N}) dp^3}{(2\pi)^3 E_i} p^{\mu_1} ... p^{\mu_n} f_0^i (1+\varepsilon f_0^i)(1+2\varepsilon f_0^i). \tag{C.12}$$

The equilibrium moments can be decomposed with respect to the flow as in Eqs. (3.46) and (3.64). The coefficients are explicitly expressed as

$$J_{kl}^{j_1,\ldots,j_N} = \frac{1}{(2l+1)!!} \times \sum_i \int \frac{(q_{J_1}\ldots q_{J_N})dp^3}{(2\pi)^3 E_i}[m_i^2 - (p\cdot u)^2]^l (p\cdot u)^{k-2l} f_0^i(1+\varepsilon f_0^i), \quad (C.13)$$

$$K_{kl}^{j_1,\ldots,j_N} = \frac{1}{(2l+1)!!} \times \sum_i \int \frac{(q_{J_1}\ldots q_{J_N})dp^3}{(2\pi)^3 E_i}[m_i^2 - (p\cdot u)^2]^l (p\cdot u)^{k-2l} f_0^i(1+\varepsilon f_0^i)(1+2\varepsilon f_0^i), \quad (C.14)$$

for $k > 2l$.

Appendix D
Thermodynamic Stability Conditions

The Landau matching conditions ensure thermodynamic stability in first order theory. If the non-equilibrium entropy current s^μ had a term proportional to Πu^μ, the derivative $\partial(u_\mu s^\mu)/\partial\Pi|_{\Pi=0}$ would remain finite. This means that the system is not in a maximum entropy state at equilibrium, i.e., it is not thermodynamically stable. The matching conditions eliminate those terms. One can explicitly show it by inserting the phase space distribution into the definition of the entropy current in relativistic kinetic theory

$$\begin{aligned} s^\mu &= -\sum_i \int \frac{g_i d^3p}{(2\pi)^3 E} p^\mu [f^i \log f^i - \varepsilon^{-1}(1+\varepsilon f^i)\log(1+\varepsilon f^i)] \\ &= s u^\mu + \sum_i \int \frac{g_i d^3p}{(2\pi)^3 E} p^\mu \delta f^i \frac{p^\beta u_\beta - q_i^J \mu_J}{T} + \mathscr{O}[(\delta f_i)^2] \\ &\simeq s u^\mu + \frac{u_\beta}{T} \sum_i \int \frac{g_i d^3p}{(2\pi)^3 E} p^\alpha p^\beta \delta f_i - \frac{\mu_J}{T} \sum_i \int \frac{q_i^J g_i d^3p}{(2\pi)^3 E} p^\alpha \delta f_i. \end{aligned} \quad \text{(D.1)}$$

Here the higher order term is dropped in the third line and the summation over the type of conserved charge J is omitted for simplicity. Then the entropy current is expressed with thermodynamic variables as

$$\begin{aligned} s^\mu &= s u^\mu + \frac{1}{T} u^\mu u_\alpha \delta T^{\alpha\beta} u_\beta + \frac{1}{T} \Delta^\mu{}_\alpha u_\beta \delta T^{\alpha\beta} - \frac{\mu_J}{T} u^\mu u_\alpha \delta N_J^\alpha - \frac{\mu_J}{T} \Delta^\mu{}_\alpha \delta N_J^\alpha \\ &= \left(s + \frac{\delta e - \mu_J \delta n_J}{T}\right) u^\mu + \frac{W^\mu - \mu_J V_J^\mu}{T}. \end{aligned} \quad \text{(D.2)}$$

By using the Landau matching conditions $\delta e = 0$ and $\delta n_J = 0$, the correction terms proportional to u^μ are eliminated as

A. Monnai, *Relativistic Dissipative Hydrodynamic Description of the Quark-Gluon Plasma*, Springer Theses, DOI: 10.1007/978-4-431-54798-3,

$$s^\mu = su^\mu + \frac{W^\mu - \mu_J V_J^\mu}{T}, \tag{D.3}$$

which only includes non-equilibrium corrections at the first order with respect to dissipative currents which are perpendicular to u^μ. This is equivalent to the expression in Refs. [7, 20]

$$s^\mu = \frac{P_0 u^\mu + T^{\mu\nu} u_\nu - \mu_J N_J^\mu}{T}. \tag{D.4}$$

Furthermore, the expansion of the entropy current at the second order reads

$$\delta s^\mu_{(2)} = -\frac{1}{2}\sum_i \int \frac{g_i d^3 p}{(2\pi)^3 E_i} p_i^\mu \frac{\delta f^{i2}}{f_0^i (1 \pm f_0^i)}. \tag{D.5}$$

The expression is negative when contracted with the flow u_μ, representing the fact that system is in maximum, rather than minimum, entropy state. This indicates that the thermodynamic stability condition is automatically satisfied at the second order in kinetic theory. Its derivative $\partial_\mu \delta s^\mu_{(2)}$ obviously requires the information of δf^i. This suggests that, unlike first order theory, second order theory in general depends on how the off-equilibrium distribution is estimated.

Appendix E
Second Order Distortion of Distribution

The expansion of the phase space distribution $f^i = [\exp(y^i) \mp 1]^{-1}$ around the local equilibrium distribution f_0^i up to the second order yields the distortion of the distribution

$$\delta f^i = -f_0^i(1 \pm f_0^i)\delta y^i + \frac{1}{2} f_0^i(1 \pm f_0^i)(1 \pm 2 f_0^i)\delta y^{i2}. \tag{E.1}$$

δy^i can be estimated in the extended moment method as $\delta y^i = p_i^\mu \sum_J q_i^J \varepsilon_\mu^J + p_i^\mu p_i^\nu \varepsilon_{\mu\nu}$, and the distortion up to the second order is expressed as

$$\begin{aligned}\delta f^i =& - f_0(1 \pm f_0)(p_i^\mu \sum_J q_i^J \varepsilon_\mu^J + p_i^\mu p_i^\nu \varepsilon_{\mu\nu}) \\ &+ \frac{1}{2} f_0(1 \pm f_0)(1 \pm 2 f_0)(p_i^\mu p_i^\nu \sum_{J,K} q_i^J q_i^K \varepsilon_\mu^J \varepsilon_\nu^K \\ &+ 2 p_i^\mu p_i^\nu p_i^\rho \sum_J q_i^J \varepsilon_\mu^J \varepsilon_{\nu\rho} + p_i^\mu p_i^\nu p_i^\rho p_i^\sigma \varepsilon_{\mu\nu}\varepsilon_{\rho\sigma}).\end{aligned} \tag{E.2}$$

In principle, the $10 + 4N$ unknowns are again determined by matching the macroscopic dissipative currents with the ones calculated in kinetic theory. Unlike in first order theory, though, the distortion involves higher order contributions in terms of the dissipative currents. The resultant ε_μ^J and $\varepsilon_{\mu\nu}$ would be, by omitting the third and higher order contributions, expressed as follows:

$$\begin{aligned}\varepsilon_\mu^J =& \left(D_\Pi^J \Pi + D_{\delta e}^J \delta e + \sum_K D_{\delta n_K}^J \delta n_K \right) u_\mu + D_W^J W_\mu + \sum_K D_{V_K}^J V_\mu^K \\ &+ \left(\delta_{\Pi\Pi}^J \Pi^2 + \delta_{\delta e \delta e}^J \delta e^2 + \sum_{K,L} \delta_{\delta n_K \delta n_L}^J \delta n_K \delta n_L \right.\end{aligned}$$

A. Monnai, *Relativistic Dissipative Hydrodynamic Description of the Quark-Gluon Plasma*, Springer Theses, DOI: 10.1007/978-4-431-54798-3,

$$+\delta^J_{\Pi\delta e}\Pi\delta e+\sum_K\delta^J_{\Pi\delta n_K}\Pi\delta n_K+\sum_K\delta^J_{\delta e\delta n_K}\delta e\delta n_K$$

$$+\delta^J_{WW}W^\nu W_\nu+\sum_K\delta^J_{WV_K}W^\nu V^K_\nu+\sum_{K,L}\delta^J_{V_K V_L}V^\nu_K V^L_\nu\Bigg)u_\mu$$

$$+\delta^J_{\Pi W}\Pi W_\mu+\delta^J_{\delta eW}\delta eW_\mu+\sum_K\delta^J_{\delta n_K W}\delta eW_\mu+\sum_K\delta^J_{\Pi V_K}\Pi V^K_\mu$$

$$+\sum_K\delta^J_{\delta eV_K}\delta eV^K_\mu+\sum_{K,L}\delta^J_{\delta n_K V_L}\delta n_K V^L_\mu+\delta^J_{W\pi}W^\nu\pi_{\mu\nu}+\sum_K\delta^J_{V_K\pi}V^\nu_K\pi_{\mu\nu}, \tag{E.3}$$

$$\varepsilon_{\mu\nu}=\Bigg(B_\Pi\Pi+B_{\delta e}\delta e+\sum_J B_{\delta n_J}\delta n_J\Bigg)\Delta_{\mu\nu}+\Bigg(\tilde{B}_\Pi\Pi+\tilde{B}_{\delta e}\delta e+\sum_J\tilde{B}_{\delta n_J}\delta n_J\Bigg)u_\mu u_\nu$$

$$+2B_W u_{(\mu}W_{\nu)}+2\sum_J B_{V_J}u_{(\mu}V^J_{\nu)}+B_\pi\pi_{\mu\nu}$$

$$+\Bigg(\beta_{\Pi\Pi}\Pi^2+\beta_{\delta e\delta e}\delta e^2+\sum_{J,K}\beta_{\delta n_J\delta n_K}\delta n_J\delta n_K$$

$$+\beta_{\Pi\delta e}\Pi\delta e+\sum_J\beta_{\Pi\delta n_J}\Pi\delta n_J+\sum_J\beta_{\delta e\delta n_J}\delta e\delta n_J$$

$$+\beta_{WW}W^\rho W_\rho+\sum_J\beta_{WV_J}W^\rho V^J_\rho+\sum_{J,K}\beta_{V_J V_K}V^\rho_J V^K_\rho+\beta_{\pi\pi}\pi^{\rho\sigma}\pi_{\rho\sigma}\Bigg)\Delta_{\mu\nu}$$

$$+\Bigg(\tilde\beta_{\Pi\Pi}\Pi^2+\tilde\beta_{\delta e\delta e}\delta e^2+\sum_{J,K}\tilde\beta_{\delta n_J\delta n_K}\delta n_J\delta n_K$$

$$+\tilde\beta_{\Pi\delta e}\Pi\delta e+\sum_J\tilde\beta_{\Pi\delta n_J}\Pi\delta n_J+\sum_J\tilde\beta_{\delta e\delta n_J}\delta e\delta n_J$$

$$+\tilde\beta_{WW}W^\rho W_\rho+\sum_J\tilde\beta_{WV_J}W^\rho V^J_\rho+\sum_{J,K}\tilde\beta_{V_J V_K}V^\rho_J V^K_\rho+\tilde\beta_{\pi\pi}\pi^{\rho\sigma}\pi_{\rho\sigma}\Bigg)u_\mu u_\nu$$

$$+2\beta_{\Pi W}\Pi W_{(\mu}u_{\nu)}+2\sum_J\beta_{\Pi V_J}\Pi V^J_{(\mu}u_{\nu)}+2\beta_{\delta eW}\delta eW_{(\mu}u_{\nu)}+2\sum_J\beta_{\delta eV_J}\delta eV^J_{(\mu}u_{\nu)}$$

$$+2\sum_J\beta_{\delta n_J W}\Pi W_{(\mu}u_{\nu)}+2\sum_{J,K}\beta_{\delta n_J V_K}\delta n_J V^K_{(\mu}u_{\nu)}+2\beta_{W\pi}W^\rho\pi_{\rho(\mu}u_{\nu)}$$

$$+2\sum_J\beta_{V_J\pi}V^\rho_J\pi_{\rho(\mu}u_{\nu)}+\beta_{\Pi\pi}\Pi\pi_{\mu\nu}+\beta_{\delta e\pi}\delta e\pi_{\mu\nu}+\sum_J\beta_{\delta n_J\pi}\delta n_J\pi_{\mu\nu}$$

$$+\bar\beta_{WW}W_\mu W_\nu+\sum_J\bar\beta_{WV_J}W_\mu V^J_\nu+\sum_{J,K}\bar\beta_{V_J V_K}V^J_\mu V^K_\nu+\bar\beta_{\pi\pi}\pi^{\ \rho}_\mu\pi_{\nu\rho}, \tag{E.4}$$

where δ's and β's are the second order distortion factors which can be calculated in kinetic theory. These new terms might lead to additional contributions on the left hand side of the second order constitutive Eqs. (3.68)–(3.71) which could be absorbed into the right hand side by using the Navier-Stokes expressions for the dissipative currents Π, W^μ, V^μ_J and $\pi^{\mu\nu}$.

Appendix F
Relaxation of Second Order Theory

Lorentz invariance is one of the most fundamental symmetry in relativity. On the other hand, since hydrodynamics is a macroscopic effective theory, it would seem that superluminal propagation does not raise a fundamental issue, especially when there appears to be no reason to assume an upper limit for c in the numerical simulation. The acausality is unfavored because if it is explicitly violated in a theory, the relativistic τ-η coordinates will not be able to properly describe the system as the superluminal events mean imaginary τ and η. Also this suggests there exists a frame where the information seem to travel back in time, which can induce unphysical instability. Note that generally speaking, causality and stability are not the same issue even though they are often related.

The second-order constitutive equations exhibit exponential relaxation of the dissipative quantities to the Navier-Stokes forms. For the analytic discussion a simplified equation of a dissipative current J_i is considered with the linear response C_{ij}, thermodynamic force X_j and relaxation time τ_{J_i},

$$DJ_i = -\frac{1}{\tau_{J_i}} \left(J_i - \sum_j C_{ij} X_j \right). \tag{F.1}$$

Here the indices i and j denote the type of currents at the same tensor order. For now the summation over j is taken and the indices are dropped for simplicity. With the Bjorken flow $u_\tau = 1$, one has

$$\partial_\tau J = -\frac{1}{\tau_J}(J - CX). \tag{F.2}$$

The analytic solution for the equation above reads

$$J = e^{-\frac{\tau}{\tau_J}} \left(\frac{1}{\tau_J} \int_{\tau_0}^{\tau} e^{\frac{\tau'}{\tau_J}} CX d\tau' + c_0 \right), \tag{F.3}$$

A. Monnai, *Relativistic Dissipative Hydrodynamic Description of the Quark-Gluon Plasma*, Springer Theses, DOI: 10.1007/978-4-431-54798-3,

where c_0 is a constant. It is related to the the value of J at the initial time τ_0 as $J_0 = e^{-\frac{\tau_0}{\tau_J}} c_0$. Note that C and X are both time-dependent. The integral is expressed as

$$\int_{\tau_0}^{\tau} e^{\frac{\tau'}{\tau_J}} CX d\tau' = \left[\tau_J e^{\frac{\tau'}{\tau_J}} CX \right]_{\tau_0}^{\tau} - \int_{\tau_0}^{\tau} e^{\frac{\tau'}{\tau_J}} \frac{\partial(CX)}{\partial \tau'} d\tau', \tag{F.4}$$

through the integration by parts. Since $X \sim \tau^{-1}$, when τ is large enough the second integral is suppressed by τ^{-1} compared with the first term. Then one obtains

$$J \sim CX + e^{-\frac{\tau-\tau_0}{\tau_J}} (J_0 - CX), \tag{F.5}$$

where the second term is exponentially damped in the time evolution. This means that the system forgets the information of the initial condition in a typical time scale of the relaxation time after the initial time $\tau \sim \tau_0 + \tau_J$ and relaxes to the Navier-Stokes value. Another interpretation of the relaxation Eq. (F.2) is that $J - CX$ approaches vanishing when the time evolution of the dissipative current $\partial_\tau J$ becomes smaller at large τ.

It is note worthy that causality is obviously not preserved if the relaxation time is too short, as the Navier-Stokes theory is recovered in the vanishing relaxation time limit $\tau_J \to 0$. The group velocity for the shear viscosity is given in Refs. [21, 22] as

$$v_{\text{shear}} = \sqrt{c_s^2 + \frac{4\eta}{3\tau_\pi (e_0 + P_0)}}, \tag{F.6}$$

which implies that the relaxation time has to satisfy $v_{\text{shear}} < 1$ and similarly for the other dissipative currents. For high-energy heavy ion collisions, the estimation with $\eta/s = 1/4\pi, c_s^2 \sim 1/3$ suggests that the minimum for the relaxation time is $\tau_\pi \sim 0.1$ fm at $T \sim 300$ MeV.

Appendix G
Transport Coefficients at the Linear Order

The determination of the transport coefficients for a given system has been one of the long standing issues even at the linear order. There are $N + 1$ scalar viscous coefficients $\zeta_{\Pi\Pi}$, $\zeta_{\Pi\delta e}$ and $\zeta_{\Pi\delta_{n_J}}$. They are generally *not* small compared to shear viscous coefficient η. However, most of field theoretical estimations state that bulk viscosity is proportional to $(\frac{1}{3} - c_s^2)$, which remains small unless the sound velocity is somehow slowed down. A clear notion on the energy deviation δe would be required for the correct understanding this "universal" smallness of bulk viscosity. The thermodynamic forces of the scalar viscous coefficients can be merged into the expansion scalar using the conservation law and the Gibbs-Duhem relation at the leading order. The bulk pressure at the first order is expressed as

$$\begin{aligned} \Pi &= -\zeta_{\Pi\Pi}\frac{1}{T}\nabla_\mu u^\mu - \zeta_{\Pi\delta_e} D\frac{1}{T} + \zeta_{\Pi\delta n_J} D\frac{\mu_J}{T} \\ &= -\left[\frac{\zeta_{\Pi\Pi}}{T} + \frac{\zeta_{\Pi\delta_e}}{T}\left(\frac{\partial P_0}{\partial e_0}\right)_{n_{J0}} + \frac{\zeta_{\Pi\delta n_J}}{T}\left(\frac{\partial P_0}{\partial n_{J0}}\right)_{e_0}\right]\nabla_\mu u^\mu + \mathcal{O}(\delta^2) \\ &\equiv -\zeta\nabla_\mu u^\mu + \mathcal{O}(\delta^2). \end{aligned} \tag{G.1}$$

It is important that the second term is not negligible as is often believed, since $\zeta_{\Pi\Pi}$ is estimated to be a third of $\zeta_{\Pi\delta e}$ with negative sign when energy-momentum tensor in a correlation function is tensor-composed in terms of the flow. A naïve ϕ^4-theory analysis [11] yields $\zeta_{\Pi\delta e} = -3\zeta_{\Pi\Pi}$ along with $\zeta_{\Pi\Pi}/T = 5\eta/3$. If the system is free of conserved charge charges, the energy-momentum conservation and the Gibbs-Duhem relation yield

$$\zeta = 5\left(\frac{1}{3} - c_s^2\right)\eta, \tag{G.2}$$

which satisfies the conjectured lower boundary $\zeta \geq 2(\frac{1}{3} - c_s^2)\eta$ in the $\mathcal{N} = 2^*$ gauge theory [23]. Since the squared sound velocity is around $1/3$ except for the crossover regions, the cancellation of the linear terms lead to small bulk viscosity.

A. Monnai, *Relativistic Dissipative Hydrodynamic Description of the Quark-Gluon Plasma*, Springer Theses, DOI: 10.1007/978-4-431-54798-3,

It implies that the existence of the Onsager cross term is a reason for the general smallness of bulk viscous coefficient in hydrodynamic systems. In other words, the diagonal bulk viscosity $\zeta_{\Pi\Pi}$ prevents the expansion of a system but the energy cross coefficient $\zeta_{\Pi\delta e}$ encourages the expansion in the effort of decreasing the temperature, leading to the overall cancellation of the effects. When the temperature is near T_c, the bulk viscosity exhibits a peak because the sound velocity slows down. There the temperature gradient in the cross term becomes small as the energy is converted to latent heat.

If one is oblivious to the existence of δe, but still performs the decomposition, the above equation has to be assumed as the definition of bulk viscosity. This would be a reason for the fact that the typical field theory calculations seem to have the universal structure of $(\frac{1}{3} - c_s^2)$ in the expression of the bulk viscous coefficient.

When the system has finite chemical potential, the expression is subject to non-trivial contribution from the density cross coefficient. For a special case where $\zeta_{\Pi\delta n_J} = [n_{J0}/(e_0 + P_0)] \times \zeta_{\Pi\delta e}$, however, the expression is dramatically simplified and one again obtains Eq. (G.2), because the sound velocity is expressed as Eq. (B.17). The sign of $\zeta_{\Pi\delta n_J}$ flips for antimatter systems, correctly capturing the fact that the overall bulk viscosity ζ is symmetric under the charge conjugation. This form of bulk viscous coefficient is employed in Chap. 5 since the main focus for that study is on the net baryon density and its dissipation.

The baryon dissipation current can be expressed in a similar fashion using the conservation laws and the thermodynamic relation as

$$\begin{aligned} V^\mu &= \kappa_V \nabla^\mu \frac{\mu_J}{T} - \kappa_{VW} \left(\nabla^\mu \frac{1}{T} + \frac{1}{T} D u^\mu \right) \\ &= \left(\kappa_V - \frac{n_{J0}}{e_0 + P_0} \kappa_{VW} \right) \nabla^\mu \frac{\mu_J}{T} + \mathcal{O}(\delta^2) \\ &\equiv \kappa \nabla^\mu \frac{\mu_J}{T} + \mathcal{O}(\delta^2), \end{aligned} \tag{G.3}$$

where κ is the effective charge conductivity. An extreme case is $\kappa_V = [n_{J0}/(e_0 + P_0)] \times \kappa_{VW}$, where the current vanishes altogether at this order because of the Soret effect. It should be noted that the vector cross terms arising from multiple conserved currents cannot be integrated in this method.

It should be noted that the hydrodynamic relations are modified by the next to leading order corrections as seen in Appendix B.

References

1. A. Monnai, T. Hirano, Effects of bulk viscosity at freezeout. Phys. Rev. C **80**, 054906 (2009)
2. K. Paech, S. Pratt, Origins of bulk viscosity at RHIC. Phys. Rev. C **74**, 014901 (2006)
3. M. Mizutani, S. Muroya, M. Namiki, Transport theory of the quark-gluon plasma based on an operator-field Langevin equation. Phys. Rev. D **37**, 3033 (1988)

4. D. Kharzeev, K. Tuchin, Bulk viscosity of QCD matter near the critical temperature. JHEP **0809**, 093 (2008)
5. F. Karsch, D. Kharzeev, K. Tuchin, Universal properties of bulk viscosity near the QCD phase transition. Phys. Lett. B **663**, 217 (2008)
6. C. Sasaki, K. Redlich, Transport coefficients near chiral phase transition. Nucl. Phys. A **832**, 62 (2010)
7. W. Israel, J.M. Stewart, Transient relativistic thermodynamics and kinetic theory. Ann. Phys. **118**, 341 (1979)
8. K. Dusling, D. Teaney, Simulating elliptic flow with viscous hydrodynamics. Phys. Rev. C **77**, 034905 (2008)
9. P. Kovtun, D.T. Son, A.O. Starinets, Viscosity in strongly interacting quantum field theories from black hole physics. Phys. Rev. Lett. **94**, 111601 (2005)
10. S. Weinberg, Entropy generation and the survival of protogalaxies in an expanding universe. Astrophys. J. **168**, 175 (1971)
11. A. Hosoya, M.a. Sakagami, M. Takao, Nonequilibrium thermodynamics in field theory: transport coefficients. Ann. Phys. **154**, 229 (1984)
12. P. Arnold, C. Dogan, G.D. Moore, The bulk viscosity of high-temperature QCD. Phys. Rev. D **74**, 085021 (2006)
13. F. Cooper, G. Frye, Comment on the single particle distribution in the hydrodynamic and statistical thermodynamic models of multiparticle production. Phys. Rev. D **10**, 186 (1974)
14. J.D. Bjorken, Highly relativistic nucleus–nucleus collisions: the central rapidity region. Phys. Rev. D **27**, 140 (1983)
15. D. Teaney, Effect of shear viscosity on spectra, elliptic flow, and Hanbury Brown-Twiss radii. Phys. Rev. C **68**, 034913 (2003)
16. T. Hirano, U.W. Heinz, D. Kharzeev, R. Lacey, Y. Nara, Hadronic dissipative effects on elliptic flow in ultrarelativistic heavy-ion collisions. Phys. Lett. B **636**, 299 (2006)
17. T. Hirano, U.W. Heinz, D. Kharzeev, R. Lacey, Y. Nara, Mass ordering of differential elliptic flow and its violation for phi mesons. Phys. Rev. C **77**, 044909 (2008)
18. T. Hirano, M. Gyulassy, Perfect fluidity of the quark gluon plasma core as seen through its dissipative hadronic corona. Nucl. Phys. A **769**, 71 (2006)
19. H. Song, U.W. Heinz, Causal viscous hydrodynamics in 2+1 dimensions for relativistic heavy-ion collisions. Phys. Rev. C **77**, 064901 (2008)
20. W. Israel, Nonstationary irreversible thermodynamics: a causal relativistic theory. Ann. Phys. **100**, 310 (1976)
21. G.S. Denicol, T. Kodama, T. Koide, P. Mota, Stability and causality in relativistic dissipative hydrodynamics. J. Phys. G **35**, 115102 (2008)
22. T. Koide, E. Nakano, T. Kodama, Shear viscosity coefficient and relaxation time of causal dissipative hydrodynamics in QCD. Phys. Rev. Lett. **103**, 052301 (2009)
23. A. Buchel, Bulk viscosity of gauge theory plasma at strong coupling. Phys. Lett. B **663**, 286 (2008)

MIX
Papier aus verantwortungsvollen Quellen
Paper from responsible sources
FSC® C105338

If you have any concerns about our products,
you can contact us on
ProductSafety@springernature.com

In case Publisher is established outside the EU,
the EU authorized representative is:
Springer Nature Customer Service Center GmbH
Europaplatz 3, 69115 Heidelberg, Germany

Printed by Libri Plureos GmbH
in Hamburg, Germany